Logic Pro 8

Logic Pro 8

Audio and Music Production

Mark Cousins

Russ Hepworth-Sawyer

ELSEVIER

AMSTERDAM • BOSTON • HEIDELBERG • LONDON • NEW YORK • OXFORD • PARIS
SAN DIEGO • SAN FRANCISCO • SINGAPORE • SYDNEY • TOKYO

Focal Press is an imprint of Elsevier

Focal Press

Focal Press is an imprint of Elsevier
Linacre House, Jordan Hill, Oxford OX2 8DP, UK
30 Corporate Drive, Suite 400, Burlington, MA 01803, USA

First published 2008

British Library Cataloguing-in-Publication Data

A catalogue record for this book is available from the British Library

Library of Congress Cataloging-in-Publication Data

A catalog record for this book is available from the Library of Congress

ISBN: 978-0-240-52047-6

For information on all Focal Press publications,
visit our website at www.focalpress.com

Typeset by Charon Tec Ltd (A Macmillan Company), Chennai, India
www.charontec.com

Printed and bound in Slovenia
08 09 10 11   10 9 8 7 6 5 4 3 2 1

Contents

Contents

Contents

Contents

About the Authors

Mark Cousins

Mark Cousins works as a composer, programmer and engineer (www.cous-ins-saunders.co.uk), as well as being senior writer for *Music Tech* Magazine. His professional work involves composing music for some of the world's largest production music companies – including Universal Publishing Production Music – with broadcaster credits including BBC1, BBC2, ITV, Channel 4, Five, BBC World and Sky One, among others. He has also had works performed by the Royal Philharmonic Orchestra, the East of England Orchestra, City of Prague Philharmonic Orchestra and the Brighton Festival Chorus, as well as having mixed several orchestral albums for BMG Zomba.

Mark has been an active contributor to *Music Tech* Magazine since issue one, responsible for the majority of cover features, as well as the magazine's regular Logic Pro coverage. As senior writer, he has also had a strong editorial input on the development of the magazine, helping it become one of the leading brands in its field.

Russ Hepworth-Sawyer

Russ Hepworth-Sawyer is a sound engineer and producer with over 13 years experience of all things audio and is a member of the Audio Engineering Society and the Music Producers Guild. Russ is currently Senior Lecturer of Music Production, Industry Partnerships Manager and Head of LCMselect at Leeds College of Music. Additionally, through MOTTOsound (www.mottosound.com) Russ works freelance in the industry as a mastering engineer, writer, educator and consultant. Russ has also contributed to Sound On Sound.

Acknowledgements

Mark's Acknowledgements

I would like to thank: my Mum and Dad for buying me a Casio CZ-3000 for my Christmas present in 1987 and starting the whole ball rolling; My wife, Hannah, and our two kids, Josie and Fred, for their support, patience, and amusement over the years, and for keeping me company in the studio from time to time; Russ, who has been a long-time and sincere friend – from our early days creating angst-ridden music, to the altogether more civilized task of writing a book! Neil Worley, for taking me on at *Music Tech* and showing me the hallowed profession of being a writer; Adam Saunders, for putting up with my lack of musical output for several months; and finally, Polly, Deb and Cath for looking after Fred in those final hectic few weeks!

Russ' Acknowledgements

I would like to thank: my wife Jackie and son Tom for putting up with my absenteeism during the writing of this book – thank you!; My Mum, Jo ("Hi Mum"); My parents-in-law Ann and John for all their help; Mark for entering into yet another crazy project with me and for being the Logic Guru and top chap he is!; Max Wilson for his endless support and forcing me to move to Logic in the first place... thanks boss!; Jenn Chubb for her support and excellent work at LCMselect; Olivia Flenley for the use of her song 'RISM' within the screenshots of some of my chapters; Jon Miller for the use of his animation file, wherever you are!; Iain Hodge and Peter Cook at London College of Music; and my colleagues and students at Leeds College of Music.

Both authors would like to thank Gorden Keppel from Apple for taking the time to technically check our words, and Catharine Steers for her support and hard work throughout the writing of this book.

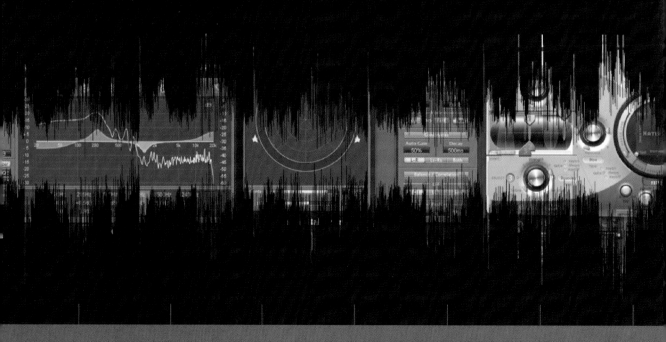

The Logic concept

1.1 Introduction

It's hard to imagine a more complete system for music and audio production than Logic Pro 8: multitrack recording and editing, a full suite of virtual instruments and effects, and a seamless workflow that takes you from the beginning of your project right through to the delivery of the final production master. Yet, with such a complete system comes the daunting task of understanding how the elements of Logic Pro 8 knit together to produce a professional-sounding result. For example, where do you begin to start writing music or making a recording in Logic? What are the virtual instruments and plug-ins used by the professionals to create release-quality output? And how can you transform those poorly performed band recordings into a polished CD?

So, let's be clear from the start, this book isn't just another instruction manual for Logic Pro 8. Instead, we've taken a process-driven approach that appraises, understands, and explores the features of Logic Pro 8 in a way that matches the structure and order of the production process. More than just a technical description of the functions of Logic Pro 8, therefore, we'll look at how the varied elements of Logic Pro 8 relate to the demands of audio and music production. With all but a few exceptions, most of the chapters focus on a specific part of the production process – whether it's initial track laying, sound design, or mastering your finished mixes to produce the final CD – highlighting the relevant parts of the application that guarantee a professional-sounding audio product. We'll also look at techniques that go beyond the scope of the manual – practices like parallel compression, for example, that many engineers use and abuse on a daily basis.

If you're starting off from scratch, it's easy to be overwhelmed by the sheer size and complexity of an application like Logic Pro 8. However, it isn't essential to know the entirety of the application to start producing music. Get to know the components that are most relevant to your way of working and build from there – use plenty of presets, Apple Loops, and so on, to get you kick started – and then enjoy the process of exploring each element that little bit further. Ultimately, Logic Pro 8 is tool that will grow with your experience – a system that will surprise at every turn and open up new possibilities whenever you want to explore the software further. With this book, you'll at least have a reference to aid you in that process, but don't be afraid to experiment to find out how Logic Pro 8 best fits into your unique creative process!

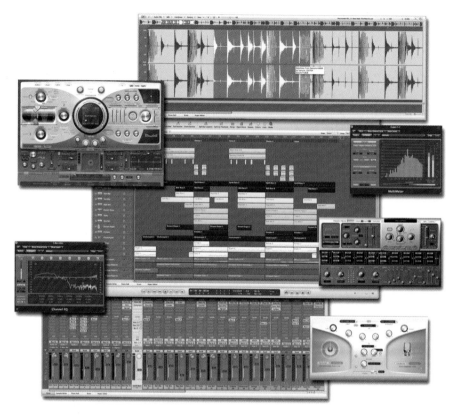

Logic Pro 8 includes an impressive array of features, but understanding how they integrate into the production process might not be immediately apparent.

1.2 A brief history of Logic Pro 8

Like the other "old-timers" of computer-based audio production – including Cubase and Pro Tools – Logic Pro 8 is an application with a rich and long heritage in the industry. Born from the ashes of C-Lab's Notator and Creator in 1993, Notator Logic (as it was then called) was an attempt to create a visual, region-based production environment for MIDI sequencing. Building blocks, or regions of MIDI data – used to control hardware synthesizers and samplers – could be arranged on the computer screen, with a clear visual representation of the structure of the arrangement. What was unique about Logic, though, was that the application was completely configurable – users could create virtual presentations of their studio, known as an environment, for example, or combine different editor windows in a completely configurable user interface.

Knowledgebase 1

Studio v. Express: the many flavours of Logic

So as to match your precise production needs and available budget, Logic is available in two principle versions – Logic Studio and Logic Express. As you'd expect, Logic Studio is the more complete package: with a range of ancillary applications including MainStage, WaveBurner, Soundtrack Pro 2, and Compressor – as well as wealth of sound content in the form of Apple Loops, EXS24 instruments, and so on. The main component, though, is the Logic Pro 8 application itself, which is the centrepiece of any music or audio production-based activity on the Mac. In effect, the additional applications build on Logic Pro 8's core functionality – with MainStage, for example, allowing you to take Logic Pro 8's instruments and effects on the road, while Soundtrack Pro 2 allows you to better integrate your work with professionals working in film and TV post-production.

Logic Express 8, on the other hand, is a more cost-effective introduction to world of music production in Logic. Although Logic Express 8 lacks the ancillary applications and full sound content of the complete Logic Studio, it does provide a feature set almost identical to that of Logic Pro 8. On the whole, the omitted features largely relate to professional applications – using TDM/DAE, for example, distributed audio processing, or surround sound mixing. The list of available plug-ins, so important to "in the box" audio production, is almost identical, with the possible exception of a few instruments like Sculpture and the EVP88, alongside the Space Designer and Delay Designer audio plug-ins.

In writing the book, therefore, we concentrated on the main features and processes applicable to both Logic Pro 8 and Logic Express 8 (henceforth referred to simply as "Logic"). Where appropriate, we have referred to some of Logic Pro 8's extra features – like 5.1 surround sound mixing, or WaveBurner – but in most cases, a Logic Express user will be able to achieve much of what this book details.

Audio functionality was added to the application in 1994, with the release of version 1.7, allowing Logic users to combine both digital audio and MIDI data all in the same arrangement (although initially, only with expensive Digidesign audio hardware). Virtual Instruments followed in 2000, making the system a complete production environment where a track could be composed, mixed, and mastered all in the one computer, and arguably, without the need for any extra third-party software. Although revolutionary at the time, this method of production has now become the norm, with many musicians and engineers largely working entirely "in the box".

Apple acquired the company that originally developed Logic – Emagic – in 2002, with its programming team joining Apple's, and Logic Pro becoming part of Apple's prized suite of media-based applications, including Final Cut Pro, Shake, and Aperture. The partnership led to many of Logic Pro's technologies migrating into other Apple applications – most notably with the introduction of GarageBand – as well as Apple making Logic Pro an increasingly more price-competitive option, with both the absorption of previously optional software components into the main application (like Space Designer, the EXS24 Sampler, and the ES2 synthesizer), and, with the release of Logic Studio, a halving of its retail price.

With Logic Pro 8, Apple have made some big moves to make the application significantly easier to use, and much more in line with the usability of their other media products. As a result, it's never been a better time for new users to join the Logic Pro 8 fold – both with respect to its affordability and the significantly easier learning curve!

1.3 Why choose Logic Pro 8?

Opinion and debate will always rage as to the "best" digital audio workstation, but there are a number of factors that give Logic Pro 8 the edge over alternative solutions. Certainly, if you're trying to make a decision between different audio applications – all with such a compelling range of features – it's well worth understanding some of their main overriding benefits, and seeing whether these align with your intended method of working.

Complete integration with Apple hardware and software

Being part of Apple, you can guarantee that Logic Pro 8 will make optimal use of both Apple's computing hardware and the operating system that ties it all together. For example, where other developers might lag behind certain OS updates, Logic Pro tends to be first off the block supporting major upgrades like OS X Leopard. On top of this, Logic Pro has always stood out from the crowd in terms of its efficient use of DSP resources, suggesting a well-coded audio engine, as well as plenty of integrated components – like the EXS24 sampler or Space Designer reverb – that ensure a completely optimal use of the processor.

Exhaustive range of plug-ins and instruments

Logic Pro 8's integral range of instruments and effects is easily the most comprehensive set available in any off-the-shelf DAW. As well as standard studio stalwarts like compression, reverb, and equalization, Logic Pro 8 includes a number of contemporary effects and plenty of software instruments covering everything from vintage Hammond organs to cutting-edge component-modelling synthesizers. Ultimately, with such a diverse collection of tools, you can easily produce a professional, release-quality output without having to resort to additional third-party plug-in and effects, although of course there's no reason why you can't add these at a later point should you wish to do so. As the old saying goes, the only limit with Logic Pro 8 is your imagination....

Effective combination of MIDI and audio editing

While some applications have strengths in a particular area (like MIDI or audio, for example), Logic Pro 8 presents an effective hybrid solution for both MIDI-based composition and audio recording and editing – which probably goes to explain why so many working composers operate completely within the realms of Logic Pro 8. Its flexibility also makes it possible to use Logic Pro 8 in a wide range of audio production environments including 5.1 and surround sound mixing, sound for film, and other multimedia applications.

Flexible audio hardware

Rather than being tied into one type of audio hardware, Logic Pro 8 users have the opportunity and flexibility to select the audio hardware for the way they want to work. For example, a simple laptop solution could just make use of the internal audio hardware, or simple two-in two-out USB interface, while professional users could use dedicated Apogee audio interfaces, or even use Logic Pro 8 as the front-end to a full Pro Tools|HD rig.

Knowledgebase 2

Installing Logic Studio

The full version of Logic Studio is now a 46 GB install, with a range of applications (including Logic Pro 8 itself) as well as five Jam Packs of Apple Loops and instruments. You'll also need to be running Mac OS X 10.4.9 or later, using at least a 1.25 GHz G4 processor, with a minimum of 1 GB of memory. If you intend to use a large amount of sample-based instruments, you might also want to consider raising your RAM resources to 2 GB or more of physical memory.

On running the installer, you'll be presented with various options for installing Logic. If you've got a large enough hard drive (150 GB+) it's well worth installing the complete package, although users with smaller drives, especially with a large amount of existing data, might want to consider a more slimmed-down install. Arguably the biggest use of disk resources therefore, are the Jam Packs, Sound Effects, and Music Beds installed as part of the Logic Studio Content. Try opening up these folders in the install dialogue box (click on the small arrows) and deactivating the elements you don't want to install, maybe deselecting genres that don't interest you so much. As you remove items from your installation requirements you'll notice the Space Required indicator reducing accordingly. As a guide, try to leave at least 50% of your drive's resources still available after installation.

On the whole, the Applications themselves (including Logic Pro, Soundtrack Pro, MainStage, and WaveBurner) don't tend to use up anywhere near as

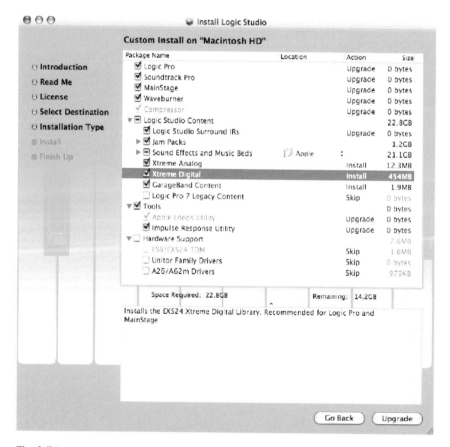

The full Logic Studio install can take up to 46 GB of drive space, although it is possible to use the custom install feature to add or remove items from your installation requirements.

much space as the sound content, allowing even the most jam-packed laptop hard drive to accommodate Logic 8. Slimmed down the minimum, you should be able to get the application-only install down to around 3 GB. If you own any legacy Emagic hardware (like the Unitor/AMT8 MIDI interfaces, or A62m Audio interfaces) do check the Hardware Support folder so as to install the latest and most up-to-date drivers for these products.

In this chapter

Knowledgebase

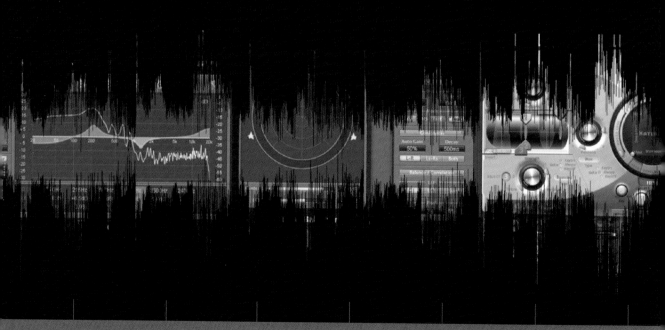

2

Logic's interface

2.1 Introduction

Negotiating the interface of any audio application is vital to understanding how it works and the precise method involved in creating a finished audio recording. Logic's interface is no exception, providing its own unique slant on the production process: with a number of editor windows, mixers, media browsers, and so on, all used to sculpt and refine your audio output. Navigate Logic's interface in an informed way, therefore, and the application will become a seamless and enjoyable part of your audio production workflow, rather than a stumbling block to your creative exploits.

In this chapter, we're going to take a look at both the overarching principles of using Logic, alongside the specific components of the Logic interface and how these integrate into the production process. From this important stepping-stone, we can then begin the process of creating your own projects and taking a more detailed look Logic's role and input in audio and music production.

2.2 What Logic can record

Logic principally works with two types of data – Audio and MIDI. Physical instruments and external sound sources are recorded directly into Logic as audio sound files. You might, for example, plug your guitar straight into a D.I input on your audio interface, or set up a number of microphones connected to your interface's mic preamps – all of which will be recorded as audio files directly into Logic's Arrange window. Once recorded, you can edit and mix these files to create your finished track, ready to be burnt on to CD.

As an alternative to audio files, you can also record MIDI data directly into Logic using an attached MIDI keyboard, which, in turn, can be used to control Logic's integral virtual instruments (or other third-party Audio Unit instruments), as well as external hardware MIDI synthesizers and samplers. Unlike audio recording, MIDI production provides an unprecedented amount of scope over both the performance and sound of the music; although it can sometimes lack the life and energy of a music performed by real musicians.

Of course, it's highly likely that most projects in Logic will use a combination of both audio and MIDI recording together to create the optimal presentation of your track. In that respect, the combined power and functionality of the Arrange window in handling such projects makes Logic a superb production solution.

2.3 The Arrange window

The Arrange window is the nerve centre of your work in Logic, providing a range of different editing and arrangement features to piece your project together. Placed in the middle of the interface, and covering the majority of the screen, is the Arrange area, displaying a list of current tracks and instruments residing in the project (in the track list down the left-hand side of the arrange area), alongside the various regions that form the structure and development of your song over time. By adding or deleting tracks, or moving regions around the screen, we can visually build up the structure and arrangement of the song accordingly.

The main arrangement area provides a visual overview of the tracks used in the project, as well as the arrangement of the song defined by the various regions.

Springing up from the sides of the Arrange window are a number of other functional "areas". These areas relate to activities involved in the production process and the tools of editing, arranging, and mixing audio and MIDI data. For example, in the initial stages of production, you might need to input the various signals to be recorded, checking their respective recording levels and making a suitable "monitor mix" – in this case, opening up the Mixer tab at the bottom of the Arrange window. Later on, you might want to import some additional audio files – using Logic's Browser – or edit the timing of a MIDI performance using the Piano Roll.

Use the other areas of the Arrange window to access more detailed application-specific editing tools and features.

The idea with Logic's Arrangement window, therefore, is that it can provide a dynamic working environment, optimized for the tasks you need to carry out at any point in time. Although it's possible to open and shut many of these extra areas using on-screen tabs, or double clicking on regions, it's well worth memorizing some of the important keyboard shortcuts so that you open them "on-the-fly". Importantly though, whatever additional area you decide to open up, you'll still have some overview of the project's arrangement, allowing you to keep a handle on how your edits affect the entirety of the track.

Knowledgebase 1

What is a Region?

One of the principle concepts of DAW system like Logic Pro is the notion of non-destructive edit. Unlike the process of editing tape – which physically

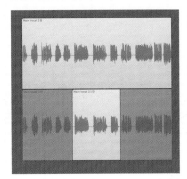

A region is simply a window on the original data, allowing you to resize the edit at any point – effectively bringing back audio or MIDI data after the original start or end point has been established.

cuts the tape into two – an edit in Logic simply changes the proportion of data played. In that respect, we need to make an important distinction between the "source file" or recording, and a "region". In effect, a region is simply a window on the original data – playing back either a small or large part of file. You can also have a number of different regions focusing on different part of the original file, but, ultimately, all sourced from the same material. Of course, all of these types of edit would be impossible to achieve with a physical, destructive editing medium like tape.

2.4 Editor areas – Mixer, Sample Editor, Piano Roll, Score and Hyper Editor

Opening from the bottom of the arrangement window, the various editor areas within Logic allow you work with and edit your project in a number of different ways.

Mixer (keyboard shortcut – X)

The Mixer balances the respective levels of the tracks included in your project, as well as instantiating different signal processing plug-ins (like reverb, compression or equalization) to process each mixer channel. As well as audio mixing, the mixer is also the primary port of call for using virtual instruments (like synthe-sizers, samplers and vintage keyboards), allowing us to create music completely within the realms of Logic. Alongside the arrange area, the mixer is undoubtedly one of the most important day-to-day production tools used in Logic.

The mixer allows us to balance the mix, process tracks using audio effect plug-ins, and instantiate virtual instruments.

Sample Editor (keyboard shortcut – W)

The Sample Editor provides detailed, sample-accurate audio editing of a given region in the arrange area. Although a large amount of audio editing can be achieved in the arrange area alone, it is the sample editor that really allows us to comfortably work with low-level details of an audio recording – precisely setting edit points, for example, or applying unique "destructive" editing techniques like reversing, silencing, or time compression/expansion.

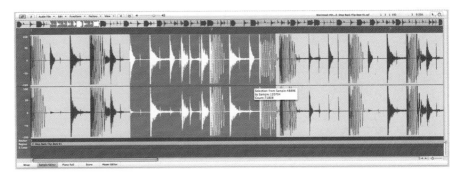

The sample editor allows you to edit and process your audio files in ways impossible to achieve in the Arrange area alone.

Piano Roll (keyboard shortcut – P)

The Piano Roll editor is the main MIDI editor in Logic, used to edit MIDI information that in turn controls virtual instruments or external hardware synthesizers and samplers. Using an intuitive display of note information based on their position on the piano keyboard (on the vertical axis) and the horizontal axis displaying both the note's starting point and duration, Piano Roll offers both unparalleled precision and ease-of-use for MIDI editing. Using the extended editing functions of the Piano Roll, you can also transform and edit MIDI information in ways that would be extremely time consuming (if not impossible) to perform by hand.

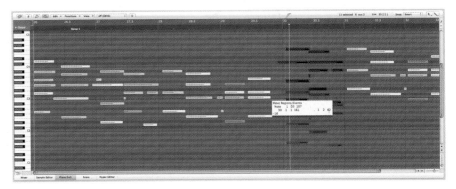

Piano Roll is Logic's main MIDI editor, with a range of editing tools and features dedicated to the precise editing of performance data.

Score (keyboard shortcut – N)

As well as providing a useful way of visualizing and editing MIDI information for musically literate users, the Score editor is the vital conduit between your raw MIDI recordings and a finished printed score readable by real musicians! Although not everybody will want to use the Score editor, it does illustrate Logic's intention to be a versatile and competent music production tool in the widest sense.

Although slightly less useful as a MIDI editor, the Score area is still a useful tool for turning your MIDI performances into a finished score.

Hyper Editor (keyboard shortcut – Y)

This final form of MIDI editor offers a more specialized solution to working with controller data – like volume or pan – as well as offering an alternative means of programming and creating drum tracks. Information is presented within a unique series of vertical beams, allowing you to quickly draw in controller sweeps, for example, or complicated drum patterns full of 16th notes. As with the score area, the Hyper Editor might not be to everyone's tastes (there are alternative ways of achieving many of its main applications), but it does form an interesting alternative for working with several different lanes of controller data outside of the Piano Roll editor.

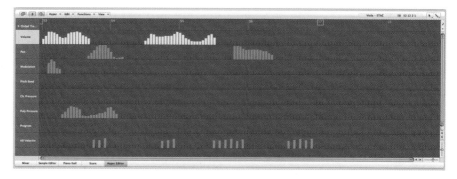

Hyper Editor displays its information as a series of unique vertical beams. As such, the Hyper Editor forms a useful solution for dealing with MIDI controller data.

To change the respective proportion of screen devoted to main arrange and the various editors, click and drag on the thin grey line between the arrange area and bottom editor area. This allows you to optimize the view based on your current priority. For example, expanding the mixer when you need full access to its compression and equalization settings, or using a minimized mixer (with just the main fader positions viewable) when you want to quickly rebalance the mix in relation to the arrangement.

Change the relative window size (using the thin grey line between the areas) to prioritize the view based on your current production activity.

2.5 Media and Lists area

Any project in Logic will be potentially comprised of a number of different audio files, plug-in settings (for instrument and effects), and Apple Loops. The Media area, therefore, is an important tool for keeping on top of these various components, and is permanently located towards the right-hand side of the Arrange window. Once open, the Media area can perform a vital role in a number of media-based activities: from the drag-and-drop importing of audio files directly into your project to the navigation of plug-in settings and the creation of sampler instruments in the EXS24.

Alongside the Media view, Logic can also use the right-hand side of the arrangement window to display a series of list-based information. Although not as glamorous or intuitive as some of the other editors, these list views are still vital to the success of the project, and the flexibility and precision of working with Logic. In total, there are four types of list available. Event presents MIDI information in its "raw" form, allowing you to see precise parameter

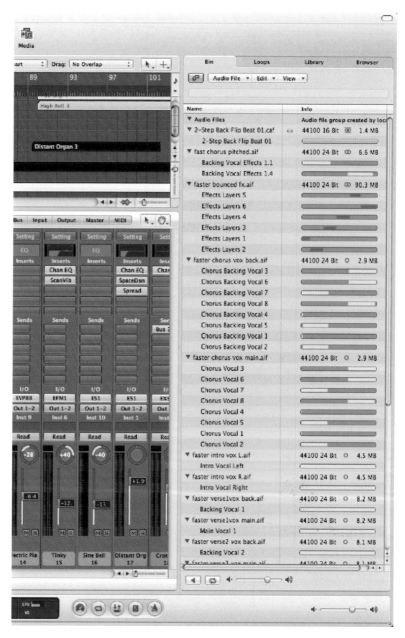

Using the Media area, you can work dynamically with the range of media files associated with a project.

values and positional data. Marker displays a list of locational reference points (like verse, chorus, middle eight, and so on), which can be an effective way of navigating long projects. The Tempo list manages a project's tempo changes, particularly vital when working to picture, while Signature offers a similar control in relation to time signature.

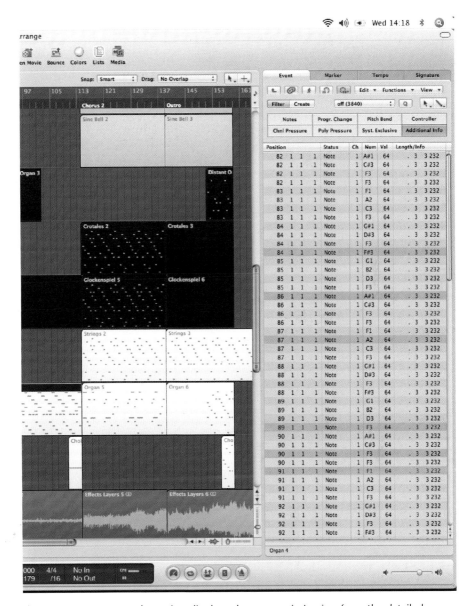

The Lists area manages the various list-based processes in Logic – from the detailed viewing and editing of MIDI information in the Event list, through to the Tempo list's set of tempo changes.

2.6 Inspector area (keyboard shortcut – I)

The Inspector area – found towards the left-hand side of the Logic's Arrange window – provides a quick-and-easy access point to key parameters relating to the regions and tracks used within your project. The Inspector is a powerful tool for audio production work in Logic, performing many essential tasks like quantizing on MIDI regions, or fades and crossfades with audio regions, as well as being an excellent way of view smaller parts of the audio mixer.

The Inspector is divided into three principle areas, which, looked at from top to bottom, are the region parameters, track parameters and the arrange channel strip (see below for more information). It is possible for both the region and track parameters to be minimized, in which case you'll need to open them (using the small arrow) to access the complete parameter set.

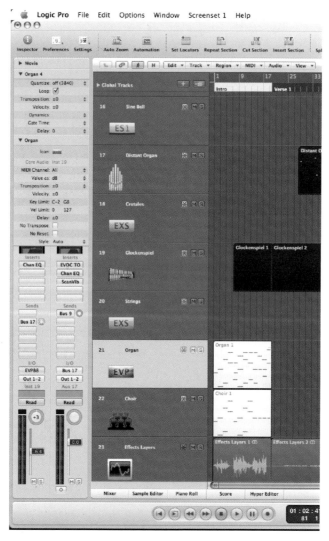

The Inspector provides a quick access point to parameters relating to audio regions and tracks, as well a track-by-track display of the audio mixer.

Region parameters

Region parameters govern the playback data of a particular audio or MIDI region in the arrangement. For example, a MIDI region can be quantized, transposed, delayed and looped, all directly from the region parameters box. An audio region, on the other hand, has its own unique set of parameters, largely in relation to the use of fades and crossfades. Besides being applied on a region-by-region basis, the region parameters box is also a powerful tool for modifying playback data across several regions at the same time – maybe applying a massed quantizing setting, for example, for all the MIDI regions in your arrangement.

Track parameters

Track parameters establish a number of parameters in relation to the track. Audio tracks tend to have slightly fewer parameters, whereas MIDI and virtual instrument tracks present slightly more control.

Arrange channel strip

Not a full mixer as such, but the inspector's channel strip view at least provides some basic access to the audio mixer on a track-by-track basis. The second channel strip illustrates the "next step" in the channel's signal path. For example, this could be the main outputs, or an aux fader, if the track was being fed to reverb or delay plug-in via a bus send.

2.7 Transport and Toolbar

Framing the extreme top and bottom of the Arrange window is the toolbar and transport bar, respectively. The toolbar offers quick access to important menu-based functions in Logic: from basic settings and preferences to various editing tools like region nudging, region splitting, automation and strip silence.

The transport bar includes the all-important transport controls (play, rewind, fast forward, and so on), a display area (including SMPTE time, CPU usage, and so on), and finally, a number of Mode buttons, to activate functions like the metronome and cycle playback.

To match your preferred way of working, it's worth noting that both these bars can be redesigned as you see fit by Ctrl-clicking on part of their greyed-out surface. In both examples, you can add further functions, and in the case of the toolbar, change the relative positioning and arrangement of the buttons.

Use to the toolbar to access features only usually available via keyboard shortcuts or the menu.

Navigate your project and invoke various mode buttons via Logic's transport bar.

2.8 Tools, local menus and contextual menus

With many of Logic's principal audio and MIDI editors, alongside the main arrangement area, you'll find series of tools, dedicated to carrying out different activities or tasks. These tools can be configured for the left- and right-hand mouse button using the small drop-down menus in the top right-hand corner of the respective area. For example, the arrange area includes tools that relate to the manipulation of regions, as well as automation data. When you're editing automation data in the arrange area, therefore, it might be beneficial to place the main pointer tool on one mouse button, with the Automation Curve tool on another.

Pressing the Escape key at any point will also bring up the particular toolbox menu for that area, allowing you to quickly switch tools without having to travel to the extremes of the area's window.

Alongside Logic's global menus, you'll also find a set of local menus unique to each area, many of which contain powerful editing features and functions that go beyond what can be achieved by mouse alone. Although not essential, it's worth familiarizing yourself with these menu features over time.

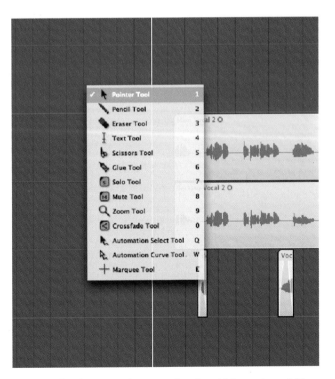

Each area has its own unique set of tools, which can be quickly accessed via the Escape key.

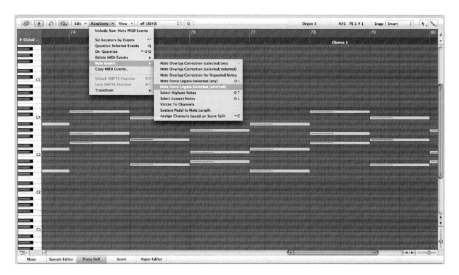

Each area will have its own local menu, with a range of functions that are unique to the activity taking place.

In many cases, it also possible to access these features via the use of contextual menus, directly activated by Ctrl-clicking in the relevant area. As you'd expect, the precise role and content of the contextual menus changes between areas, but you'll also find that the menu also adapts given the precise

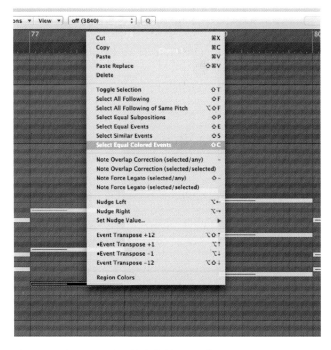

Contextual menus – by Ctrl-clicking on the screen – access many features that are unique to the area or object you click on.

23

part of the screen or objects you click on. For example, in the arrange area, the contextual menu will differentiate between ctrl-clicking in the main region area, regions themselves, the bar ruler, and the track list, with different functions appearing each time. Although bewildering at first, as you get more used to these features, you'll find the contextual menus a powerful and immediate means of accessing the real depths of Logic's editing prowess.

2.9 Adjusting how you view the arrangement: zooming in and out

Varying the magnification of your project is essential both in terms of keeping an overview of the "macro" arrangement of your track, but also, to be able to hone in on certain details, even down to sample-level accuracy. In the arrangement area, the two zoom controls (for the horizontal and vertical axes, respectively) can be found towards the bottom right-hand corner of the window – note that can also use Ctrl+ any of the arrow keys to achieve the same movement in an out of magnification. You can also use the zoom tool to drag-enclose a specific portion of the screen you want to view, with a double click taking you back to your original magnification level.

As an alternative to zooming in and out, you can also use the menu option View > Auto Track Zoom (keyboard shortcut Z) to enable a track-dependent horizontal zoom. This allows you to better view the track that you're currently working on, although, of course, this is only in relation to the horizontal rather than vertical axis.

Using the various zoom functions allow you to understand both the overall structure of your arrangement and the smallest details within it.

Another type of magnification is the waveform intensity. This effectively changes the volume scaling of the waveform display, allowing you to better see quieter, more discrete parts of the waveform that you usually wouldn't be able to see. To expand the waveform, simply click and hold on the waveform icon next to the horizontal/vertical zoom controls, and use the slider to scale accordingly. Although the waveform might end up looking distorted, the meters should, in fact, confirm that it isn't.

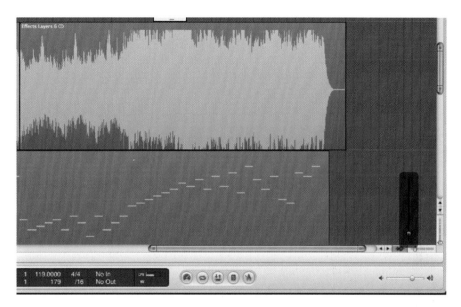

Changing the scaling of the waveform will allow you to improve your audio editing in relation to quieter parts of the signal.

In this chapter

Knowledgebases

Walkthrough

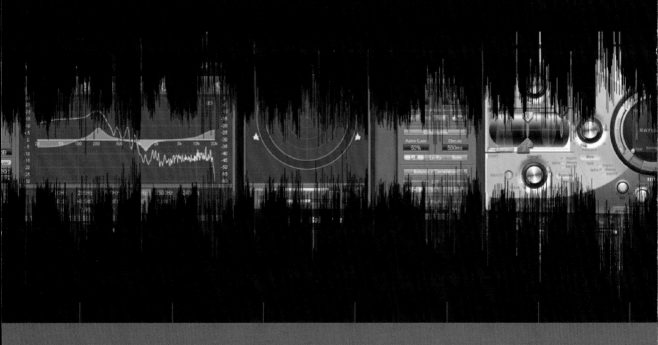

Getting connected

3.1 Introduction

The traditional model of the recording studio included, amongst many other things, a mixing console, multitrack tape recorder, outboard processing and monitoring. Understanding how the studio connects together is crucial knowledge to the recording engineer. Although this model has to some degree been swallowed up within Logic, many connections will still be necessary. Knowing what connections you need and how to get them to work is all part of working on a production.

Whilst the Mac's onboard audio interface is very respectable, it comes in a variety of differing configurations depending on the computer you own. For example, most Apple laptops typically contain a set of stereo speakers and a headphone jack for the output as standard. For the input, these machines are likely to contain a mono electret condenser microphone built into the screen casing. Whatever the onboard input, at some point you might need to bring on stream an audio interface to possibly connect either a condenser microphone using phantom power or multiple inputs. Additionally, it is likely you will wish to connect an external master keyboard for performing and programming your MIDI regions.

In this chapter, we look at the ways in which Logic communicates and connects to the outside world via the computer. We explore both MIDI and Audio communication systems and the variety of ways in which this can be achieved, by selecting the appropriate devices and paying attention to how to get the best out of them. Although MIDI and Audio are the main two conduits for our work in Logic, other aspects such as hard drives, timecode, control surfaces, ProTools connectivity and ReWire all need consideration for the fully fledged Logic User.

3.2 The studio environment

To get Logic working at its best, we need to understand the studio environment in which Logic is likely to sit. The music studio has indeed changed over the years, and for most Logic users, the use of outboard equipment such as mixers, compressors and other processors is likely to be minimal as most work will be done within the Mac. However, the computer will still need to be connected to various elements of your studio environment such as MIDI interfaces, Audio interfaces, and loudspeakers (or monitors as we call them in the audio business).

Audio considerations

Starting a project using Apple Loops and the onboard virtual instruments will yield some excellent results, and can be done with the internal audio facilities straight out of the box. However, at some point, you may want to record a vocal or a guitar track as an overdub and realize that some expansion is necessary. In this expansion it is likely that your setup will need to accommodate new inputs and ways of working, and as such your choice of audio interface will be important and is something we'll consider later in this chapter.

There are broadly three considerations at this stage. First is whether you simply need a few decent inputs for stereo pair work and overdubs. Second is whether Logic will be your only mixer and integrate your outboard equipment. Last is if you are planning to use an external console to mix your music together.

1. *Basic setups.* In the first instance, a simple interface is all that is required with a range of inputs and outputs to suit the work you're undertaking. It is presumed in such a setup that Logic will undertake the work of the mixer and the interface's outputs will simply need to be connected to your hi-fi or powered monitors.
2. *Mixing internally.* If you're likely to integrate your outboard equipment into the Logic environment, using only the internal mixer, then you should consider an interface that has the required number of inputs and outputs to connect to the outboard equipment. For example, you may wish to integrate your favourite analogue compressor and effects unit into the system at the same time as benefit from the sheer power and automation features of Logic's mixer. In this instance each external device will require up to two outputs and inputs. Therefore, a 4-input and 6-output interface will be required, which includes two outputs used for monitoring your mix.
3. *Mixing out of the box.* Integrating Logic into a studio with an established mixer is becoming a popular approach for some professionals wishing to gain the benefit of analogue summing. To do this we will require a different interface system that accepts many inputs and outputs.

Choosing an audio interface will depend on these considerations and the specifications you require. If you find yourself working alone on overdubs such as the odd vocal or guitar track then a USB-based interface will be more than ample. However, for those larger recording projects where large numbers of inputs are required such as drums or a full band, a larger horse-powered device probably using FireWire or PCI will be required. These come in many different forms for different applications and are discussed in this chapter.

Before investing in your audio interface, consider the studio environment in which you work, or aspire to work in the near future. Choose the most appropriate device that does what you want it to do with the musical activities you hope to be involved in and the equipment you have or hope to own soon.

MIDI connectivity

When early versions of Logic, and its predecessor C-Lab, were the norm, the concept of audio integration and a complete music production environment was a far-flung dream. These then were purely MIDI sequencers, moving and editing digital representations of the notes played against time. The cabling was somewhat different in those days with MIDI cables transmitting data from the computer through to sound modules and samplers. They would in turn have their audio signal connected to a mixer so that the signals could be monitored.

MIDI modules and keyboards are still used and if you're connecting these devices to Logic you need to invest in a MIDI interface, which we'll discuss later. It is, however, becoming increasingly unlikely that you will be using the 5-pin DIN format connector, which MIDI originally adopted. These days many master keyboards and sound modules can be connected directly to your computer using the Universal Serial Bus (USB). As such the MIDI connection protocol is now sometimes resigned to being the "invisible" language.

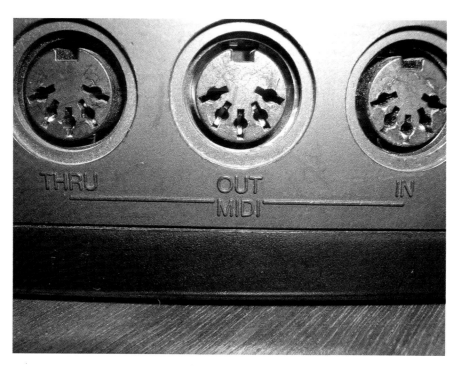

The conventional MDI socket uses a 5-pin DIN format.

MIDI sockets usually come labelled in three ways: IN, OUT, and THRU. As you might imagine, MIDI IN receives instructions from another device, for example, the sound module will receive an instruction from the MIDI OUT of Logic's MIDI interface. If the sound module sends information back to Logic such as MIDI controller data then the MIDI OUT of this will need to be connected to the MIDI IN of the interface. These MIDI connections apply to every MIDI-compatible device within your studio.

Additionally, MIDI THRU is included in the MIDI protocol to allow for something that is known as Daisy Chaining. In the early days of MIDI, it was unlikely that your sequencer would have more than one MIDI output, two at very most with C-Lab Notator on the Atari Computer! As such, if you had more than one MIDI device, it was important that you could connect it to the sequencer with ease. Enter the MIDI THRU socket which replicated the MIDI IN data and passed it through the device allowing you to connect it straight to the MIDI IN of the next device unaltered. On rare occasions some MIDI devices would merge both the MIDI OUT and MIDI THRU of a device so that a continuous loop back to the sequencer could be made. In this day and age, multiples of MIDI INs and OUTs are possible using modern MIDI interfaces; therefore, the MIDI THRU socket is less used.

It is really important to appreciate MIDI as a communication medium, as this is the language we use when programming virtual instruments in Logic. As you program up anything within the Piano Roll Editor, the language behind

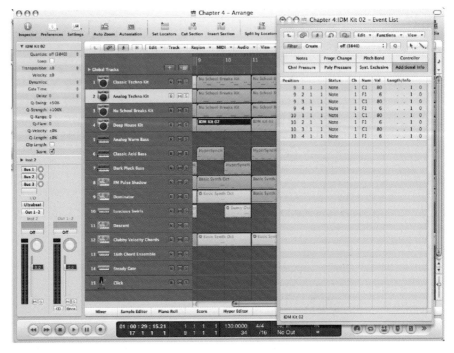

The MIDI information can be clearly seen within an Events List (Windows > Events List or Apple + 0). The Events List shows the corresponding MIDI information contained within the selected MIDI region.

those notes and commands is simply MIDI. To see more information on these notes, call up the Events List (Window > Event List or Apple + 0) and look at the information it gives as in the example below. We'll discuss MIDI in more depth in Chapter 6.

3.3 USB devices

Connecting devices to the Mac is now a much easier affair with USB than in the earlier years. USB has enabled many different devices including mice, keyboards, scanners and of course audio interfaces to be connected to the computer using the same connection protocol. USB 1.0 came as a slow proto-col of 12 megabits per second (Mb/s), which would slow with too much traf-fic placed upon it with multiple connections using a USB hub. USB 2.0 came along a few years later, and offers a considerably faster connection speed of 480 Mb/s allowing the USB to be fast enough for hard drives and audio inter-faces with higher track counts and higher data rates.

Most external devices and peripherals for computers these days are connected using the USB, including most, if not all used for music making. USB is a flex-ible system which enables multiple devices to be connected to one computer with the use of a USB hub. Each hub acts as a splitter/merger box and typically allows for between 4 and 10 devices to be connected at once. The USB limit is much larger than this, but is unlikely to be exceeded by most studio users.

In this section we'll explore some of those USB peripherals you're likely to con-nect to your Mac when working with Logic.

USB MIDI

To connect up MIDI, an interface is required which enables some form of trans-lation from USB to MIDI. This is usually done in one of two ways. The first of which is via a dedicated USB interface such as the Fastlane-USB 2-by-2 MIDI interface developed by Mark of the Unicorn (MOTU) shown below. From one USB lead, two MIDI outputs and two MIDI inputs can be connected. These con-nections would be sufficient for a small studio using mainly internal instruments to Logic with an external MIDI keyboard. By way of expansion, another synthe-sizer could be connected at a later date to expand your arsenal.

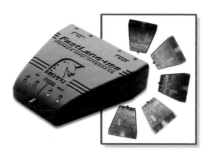

Mark of the Unicorn's Fastlane is a solid and inexpensive USB MIDI interface offering two INs and OUTs. (Image courtesy of Mark of the Unicorn).

For larger MIDI-based setups incorporating a lot of separate units including samplers, sound modules, keyboards, etc., units such as the USB-connected MOTU MIDI Time Piece (shown below) can be used, which offers eight MIDI inputs and eight MIDI outputs. This is a top-of-the-range interface which does a lot more than simply connect MIDI to the computer; it allows patching between MIDI inputs and outputs, merging MIDI data where needed, and synchronization with external digital devices such as the ADAT.

Setting up MIDI interfaces should be automatic provided they correspond with Apple's Core MIDI. Core MIDI was written by Apple to ensure that all

MOTU's MIDI Timepiece is a veritable classic and has been the mainstay of MIDI interfaces for high-end studios. It is not only a MIDI interface but also a MIDI router amongst many other features. (Image courtesy of Mark of the Unicorn).

services within the operating system operate with any MIDI interface connected. Provided your interface is understood by OS X, this should automatically be recognized and picked up within Logic.

USB audio interfaces

Audio interfaces come in many forms, but typically USB-based devices are self powered which handle two inputs and two outputs. In some instances there may be additional MIDI sockets found here too. These usually inexpensive units may not always offer the highest grade sound quality, lowest latency (see knowledgebase), or the largest range of sample frequencies and bit depths, but they are excellent solutions for where portability and flexibility are the key. Ideal uses for this interface are simple stereo-pair recordings and overdubs. A large-scale live recording would require a larger interface with a more comprehensive I/O count, but to track up a large work could be easily built up with this type of interface using overdub techniques.

There are a huge range of USB audio interfaces available and choosing the right one can be difficult. Interfaces such as these vary in price, depending on sound quality and features. Manufacturers such as M-Audio and Edirol make a wide range of excellent inexpensive interfaces for this purpose such as the UA25 shown below. Edirol is a division of Roland, and as such this unit typically is well built and has lots of features such as a high sample rate of 96 kHz and a 24 bits bit depth, internal limiter, optical digital inputs and outputs, and MIDI IN and OUT.

As mentioned earlier, USB version 1.0, which is rarely used on new devices these days, was a very slow connection and many of the stereo interfaces

The Edirol UA25 by Roland is an inexpensive audio interface with an impressive number of features for musicians on the move. (Image courtesy of Roland Corporation).

were restricted to 44.1 kHz and 16 bit. Most devices are now USB2 and have a data rate of 480 Mb/s, which is similar to that of FireWire 400. Therefore, whilst in theory any USB audio interface could handle a large input/output count (I/O count), there would be interruptions for other devices connected to the computer; thus, it is less reliable. This is due to the data throughput not being as stable as FireWire at present. As such FireWire tends to be adopted for larger audio requirements.

3.4 FireWire devices

In the early days of USB, it was only possible to connect an interface carrying a small number of audio channels as the bandwidth was limited. Therefore, FireWire was adopted to handle higher channel counts, which also added the advantage of being able to embrace higher bit rates and sample rates accordingly.

Many popular FireWire interfaces are now in use, which range from the inexpensive Mackie Onyx Satellite (shown overleaf) with two inputs and up to six outputs to more expensive units with a higher count of inputs or outputs. These devices are ideal for the larger projects you might wish to get involved in. Typically these devices offer somewhere between 4 and 8 microphone inputs and up to 10 or sometimes more analogue outputs. In addition to this there are often a multitude of digital connections, such as the ADAT lightpipe standard which can add another eight inputs and outputs to the interface.

Apogee is a well-respected high-end audio converter manufacturer who is now working closely with Apple. As a result of this cooperation two new high-end audio devices have been released. These Macintosh-only audio interfaces contain some excellent and specific benefits for Logic users. These units

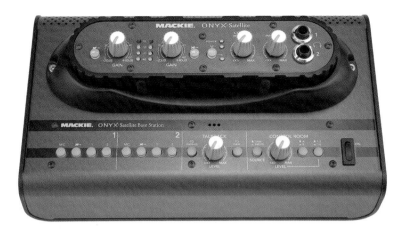

Mackie's Onyx series of preamps and audio interfaces begin with the Satellite Model, which integrates a docking station for working at home in the studio, but can be removed to work on location. (Image courtesy of Loud Technologies Inc.).

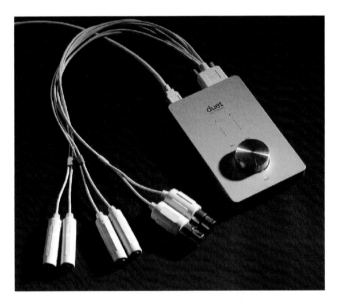

Apogee has teamed up with Apple to produce some excellent devices. The Duet is an entry device of two inputs and two outputs including a "precision digital encoder" or rotary knob for assignment to controls within Logic. The larger rack-mountable Ensemble is a larger device offering up to 36 inputs to the Mac. (Images courtesy of Apogee Electronics Corporation).

contain Apogee's reputed high-quality converters, which after all are a key part to the sound quality of your studio setup.

The first is the Ensemble which boasts eight analogue outputs and eight audio inputs, four of which are very low-noise, digitally controlled microphone pre-amps, two of which contain insert points for connection of an outboard dynamics processor and the other two also allow for the connection of high-impedance instrument connections. The Ensemble has extra facilities for digital connectivity using ADAT lightpipe to your studio setup which extends the I/O count to 16. This method of expansion is typical with many FireWire interfaces and if paired with a digitally connected console or multiple digital-to-analogue converter, such as Focusrite's OctoPre, the most can be made of the units full input and output count.

Adding additional converters such as the Focusrite OctoPre to FireWire interfaces can increase the possible input and output count. (Image courtesy of Focusrite Audio Engineering Limited).

The second device is the smaller and more portable Duet, which is essentially a two-input/two-output portable device. The accompanying break-out cable allows for six connections: two XLR mic inputs, two jack instrument inputs, and two line-level outputs. The Duet also comes with Apogee's Maestro software.

Apogee provides its own software interface between Mac OS X's Core Audio and their hardware called Maestro. This is comprehensive and offers the user the ability to configure multiple headphone mixes and outputs plus many

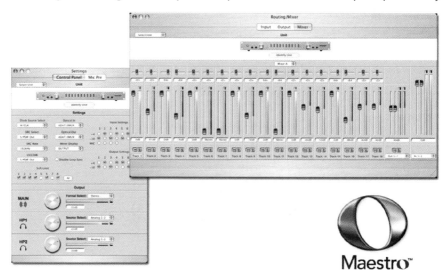

Apogee's Maestro software allows for some pretty impressive control over the inputs and outputs of your studio. (Image courtesy of Apogee Electronics Corporation).

clocking options and other intricate and useful controls. The Ensemble digitally controls its microphone pre-amplifiers and this can be managed by a
mouse using the software, but ingeniously, Apogee has also included a knob
on both interfaces, which they call a "precision digital encoder" for the purposes of controlling faders and knobs on screen.

Of course there are a plethora of FireWire devices on the market for use with
Logic which all should work with Apple Core Audio. However, before taking
the plunge, it might be worth checking on the manufacturer's website that
the device you are considering is compatible with Core Audio and Logic.

Hard drives

Recording audio is a data-intensive exercise and often as modern track counts
are appearing to increase, the internal, original hard drive of your Mac may start
to strain under the pressure. As such many Logic users either install another
hard drive in their MacPro or attach an external unit using the FireWire port to
run the audio off, thus leaving the internal drive for system requirements.

FireWire devices come in two flavours: 400 Mb/s or 800 Mb/s. Most devices
still utilize the common FireWire 400 (FW400) format due to its popularity
and reliability. Some hard drives and devices can utilize the FW800 protocol,
but it is generally rare. Although USB 2.0, rated at 480 Mb/s, is faster than
FireWire 400, its structure and ability to hub mean that it can share valuable
bandwidth with other devices such as mice and keyboards, cameras and MP3
players. As a result it has become less reliable for data efficiency, throughput,
and stability. Thus, FW400 appears to remain the de-facto standard despite it
not being the fastest presently available.

The LaCie d2 for many has become the
industry-standard hard drive of choice,
offering both reliability and flexibility
(including a rack mount available from
LaCie). (Image courtesy of LaCie).

Hard drive manufacturers such as LaCie make a series of drives suitable for the Logic user from portable, self-powering drives such as their Porsche range through to the industry standard d2. The d2 series has been the Logic professional's choice in the main part and now supports a multitude of interface formats, usually USB 2.0, FW40, FW800, and the new 1.5 Gb/s eSATA connection all on the same unit.

3.5 PCI Express and ExpressCard

Perhaps you have an existing studio with an analogue console, or are taking the plunge to replace another multitrack system with Logic. Either way, to obtain a truly large set of inputs and outputs, a top-flight system will be required using a MacPro's PCI Express (PCI-e) Slots.

These types of systems bypass the FireWire and USB connections and instead connect to a PCI-e circuit board which sits within your computer and connects directly to the computer's central processing unit (CPU). This enables much faster data rates than FireWire or USB 2.0, therefore offering larger track counts, higher sample rates, and larger word counts.

There are a few good examples of this model. One is the MOTU HD192 system shown below, which can be expanded to accept 48 inputs and 48 outputs concurrently and works with a range of softwares. Another popular PCI-based system is Digidesign's HD system using ProTools. Logic can

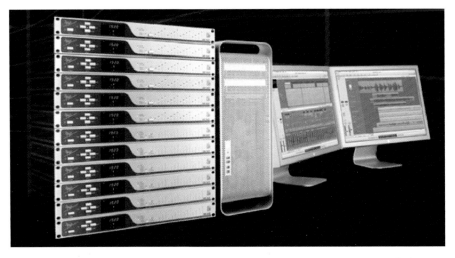

Apogee's Symphony Series connect to the Mac using the PCI-e connections and offer high sample rates and are highly expandable. (Image courtesy of Apogee Electronics Corporation).

make use of this hardware through the use of its translation software ESB TDM, which we explain further in Chapter 11.

For Laptop users, ExpressCard is the equivalent of a PCI-e card and is compatible with the MacBook Pro. This allows for high-quality interfaces and other devices to be attached to your laptop to keep you working. Manufacturers such as Apogee, with their Symphony interface and RME with the Multiface II interface, create cards to allow appropriate connections to these devices. Alternatively, the ExpressCard slot can be utilized to connect DSP expansion systems such as those offered by Universal Audio which are discussed later.

SATA and Internal Hard Drives

SATA or Serial Advanced Technology Attachment is the connection protocol between a computer and its internal hard drives. This is a very fast method of connection working at speeds of either 1.5 Gigabits per second (1.5 Gbit/s) or the newer 3 Gigabits per second (3 Gbit/s). When working in a Mac Pro, it is possible to connect up to 3 additional hard drives to that of your system disk. For audio and multimedia this is an excellent way to work as different drives can be used for differing projects, or used for backup purposes. Separating out the system drive containing Mac OS X and Logic from the audio allows the drives to be dedicated to specific tasks. However, the simple overriding benefit is the speed of SATA drives for working with high data rates such as multitrack audio.

Internal SATA drives over both speed and the ability to separate out the projects from the operating system and applications.

3.6 Audio communication and drivers

Most audio interfaces are handled by Apple's own Core Audio infrastructure. Core Audio is OS X's code specifically designed for handling sound integration on the Mac. Core Audio was designed to set a standard for audio management on OS X, hence preventing the need for each manufacturer to write interfacing code. However, many manufacturers have created accompanying software to sit beside Core Audio offering some interesting and useful features. Companies such as Apogee, MOTU, and M-Audio have taken the step of introducing a control panel which allows monitoring options, sub-mixing and even onboard digital signal processing (DSP) without having to do this necessarily through Logic. The monitoring option is a common feature and allows you to hear a direct feed of what you are recording before it is processed through Logic. Monitoring from Logic can cause small delays in the signal which can put off a performer and cause a less than ideal environment for a good take. This delay is known as latency and is covered in the accompanying knowledgebase.

Knowledgebase 1

Latency and Monitoring

When recording was in the analogue domain, the signals fed through the mixing console and open-reel multitrack tape recorder would only be delayed by an acceptable millisecond or so as the signal passed through the cables. To all intents and purposes, this was perceived as pretty much instantaneous. As this delay increases performers can be put off their performances and they might appear to be out of time.

This delay in modern digital audio is called Latency and is a common side effect of working with digital audio workstations such as Logic. Latency is caused by a delay experienced when your computer and audio or MIDI interface processes information. A common example of this is the delay between pressing the MIDI keyboard and the sound generated from an audio instrument in Logic.

Overcoming latency can often be tricky as it depends on the speed of your audio interface and computer. Monitoring is one major issue to overcome and this can be achieved by employing an external mixer to handle your headphone mix. Many inexpensive USB interfaces combat latency by employing a 'mix' knob on the unit, such as Digidesign's MBox. This allows you to blend the input signal with the output of the computer, which allows you to hear your input signal prior to any incurred latency. Firewire Audio Interfaces come with their own software-based audio mixer which performs this monitoring duty (such as MOTU's CueMix shown below) as well as often integrating on-board digital signal processing. If one of these solutions is used, then it is desirable to switch off the Software Monitoring checkbox in Preferences > Audio. Another solution is to employ Logic's Low Latency Mode.

Latency can be problematic when performing using a virtual instrument or monitoring a recording. Logic has created a neat solution called Low

MOTU's CueMix is an excellent example of
the additional mixing features that can be
provided specifically for the audio interfaces.
(Image courtesy of Mark of the Union).

Latency Mode which intelligently bypasses active plug-ins on the channel
you're working on thus freeing up processing power. This allows for the
latency to be managed to the pre-set limit of delay you have chosen in the
preferences. To change this limit go to Logic Pro > Preferences > Audio >
General. Turning on Low Latency Mode is easy as there is an icon directly
on the transport bar. In most cases this will alter the sound of the channel
you're working on, but this is only temporary for the time this feature is
engaged.

One of the best current solutions to overcome latency with Logic is to use
Apogee's Symphony hardware connected via PCI-e to the Mac. This boasts
some of the lowest latency available at the present.

Core Audio means that when you launch Logic, the connected audio interface should appear in the audio preferences. In fact, Core Audio and Logic
allow for an interface to be connected to the system mid-session, or plug-and-
play as it is known. Logic alerts you of a new interface and asks you whether
you would like to use it as in the example below. To check on the interfaces
and settings simply go to Logic Pro > Preferences > Audio where the Devices
tab should be present. Here there are three sub-tabs called "Core audio,"
"DAE," and "Direct TDM." Core Audio will relate to the device you have just

When a new audio interface is added,
Logic will notify you and ask you
whether you want to use this device. If
this does not appear, then it might be
worth obtaining the latest drivers for the
interface from the maker's website.

connected and is the first tab open to you. The other two relate to Digidesign Hardware, whose integration with Logic we'll touch on in a bit.

Looking at the Core Audio tab, there are many parameters which can be altered to get the best from your audio interface(s) called Devices in Logic. The Devices list will list all the Core Audio compatible devices available to the system at the time. Simply select the one you wish to use and hey presto you should be up and running. Core Audio makes it that simple.

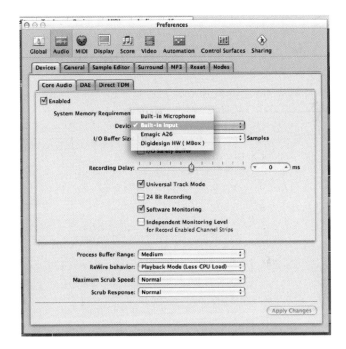

Selecting Core Audio devices is a simple process from the preferences pane.

Getting the best from your device will usually mean managing the latency you experience. As we've discussed in the Latency and Monitoring Knowledgebase there are workarounds regarding monitoring. However, the computer's driver and interface will also require time to "buffer" the information together before sending it out. The Input/Output buffer size (I/O buffer size) is measured in samples and this therefore relates to the amount of time the computer and device take before you hear the sound. In an ideal world, we'd want this buffer to be at its lowest setting of 32 samples, which at 44.1 kHz would be considerably less than a millisecond and a perfectly acceptable delay. Doing this will require a very powerful computer as more of the processing power is needed and may result in your audio dropping out or distorting. The ideal is to find a happy medium where the delay is not too long and the computer's processing power enables unhindered audio.

If latency causes problems with your recordings you can use the Recording Delay feature to move the audio back or forth so that it fits with the rest of your music. The default for this is 0 ms but can be moved by 5000 ms either way.

Bit rates in digital audio refer to how fine the measurement of the amplitude, or loudness, of the audio signal is. Compact disc (CD) recordings are presented at 16 bit and were considered resolute enough for the domestic market. However, within music production, 16-bit is considered a little too coarse to the ear and as such 24-bit is now desirable, as it measures the amplitude scale in finer detail. Therefore by increasing the detail and accuracy of the waveform, it is more representative of the original audio. To record using 24 bit, select the 24 Bit Recording check box within this pane. Remember, though, that your audio interface will need to respond to 24 bit to benefit.

Higher bit rates than 16-bit are desirable, but come at a slight price. The data this occupies on your disk is larger than that of plain 16 bit and as such you should consider whether you have enough hard drive space before the session. Roughly, you'll need nearly twice as much hard drive space as with 16 bits.

The Independent Monitoring Level feature "for Record Enabled Channel Strips" is very useful while recording as it gives the user control over the monitor level for that channel's input when in record mode. For instance, if you were to record a guitar amp close to you, it would be less necessary to hear a version of it from Logic in your headphones. Simply turn it down when in record mode and track away.

The Software Monitoring checkbox will usually be enabled as Logic presumes that you will be doing most, if not all, your audio routing and management inside as in the example above. However, if you wish to use your studio's mixer or your audio interface's monitoring features, then this should be unchecked.

The Process Buffer Range refers to the amount of time you will allow the computer to gather and process information into the buffer before playing it. The lower the range, the faster the computer will respond and the lower the latency. However, the computer may be unable to process all the required information in the time specified and drop out altogether.

ReWire, which is covered later in this chapter, is like an internal audio and MIDI patch lead connection between two pieces of software such as Reason and Logic. ReWire Behaviour refers to the type of work you are doing between Logic and the ReWired device and hence the burden on the computer. The choices are between playback and live operation. Playback assumes that the computer is being played and can use the buffer to its best effect to ensure timely playback of instruments and this process requires less CPU load. However, live mode assumes that you are playing the instrument live and as such require the connected device to be more timely when producing the sound; the computer needs to prioritize this and therefore more CPU load is used.

Knowledgebase 2

Audio in the Audio MIDI Setup

The Mac's operating system should be able to talk to your choice of interface for any audio or MIDI application such as iTunes, Reason, Live, or ProTools. The system-wide management of these interfaces is achieved in the Audio MIDI Setup which can be accessed in the Utilities menu (Apple + Shift + U from the Finder). Providing Core Audio has detected your audio interface; it should not be necessary for you to visit this utility to work straight in Logic.

When launched, the Audio MIDI Setup has two tabs, one for Audio Devices and the other for MIDI. You are likely to read here that your Default Input to be Apple's "Built-In Audio" or "Built-In Microphone". Your Default Output may be the Built-In Output; however, both this and the input can be altered to the device of your choice. Here you will note another menu entitled System Output. This latter output is offered so that the user can separate system instructions such as alerts (referred to as Sound Effects in Mac OS X's System Preferences panel) from a loud mix from being blasted through your main monitors. Nothing worse than a loud sound effect such as "Ping" clouding your 2 track master as you brush past an unexpected key!

The lower section of the panel is governed by the menu labelled "Properties For:". When you've selected the appropriate audio device you will have the ability to edit the input levels with the ability to mute these individually and also send them through to the outputs if required. The outputs can also be muted if necessary. Other editable parameters include altering the sample rate and bit depth, which will also alter the track count for some interfaces. There is even the allowance to Configure Speakers for use with multichannel setup such as 5.1.

3.7 Control surfaces

The studio traditionally was a tactile environment where physical faders, tape machines, cables and instruments were the norm. As the digital revolution has taken its hold, so much of the studio has been pared down to the one computer screen. As such the notion of grabbing the fader for any channel to make an on-the-fly adjustment becomes a little bit more involved, usually requiring several mouse clicks with some keyboard presses perhaps. Additionally computing power and software development has meant that the amount of features controlled from one mouse and keyboard has grown exponentially.

A common solution to this problem is to bring back the tactile aspect of the studio in the form of external control surfaces. These come in many forms from the one fader such as the Presonus Faderport to the fully fledged Mackie Control Universal. What is common is that most come in the form of a bank of faders with some control for other aspects of the channel strip such as pan and auxiliary sends. As you navigate around the controls, Logic should also follow suit and vice versa, giving you tactile control over your Logic mix.

Control surfaces such as Mackie's original Control Universal connected using the normal MIDI cables, although the newer version (shown opposite) and other systems, such as the FaderPort adopt the USB protocol. Other controllers are combined with audio interfaces as one package such as Roland's SI 24, thus replicating the function of the mixing console in its entirety. These and many other control surface devices are either connected to the computer using USB, FireWire, and on rare occasions, Ethernet.

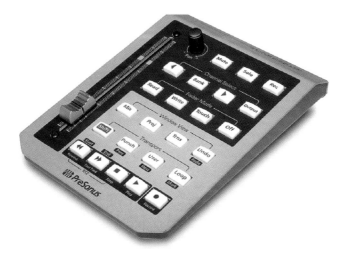

PreSonus' Faderport is a simple and intuitive control surface allowing adjustment of one fader at one time, plus the usual transport controls. (Image courtesy of PreSonus).

Mackie's Control Universal has been the choice of Logic users due to its impressive integration to Logic and its overall flexibility. (Image courtesy of Loud Technologies Inc.).

Setting up a control surface is easy, provided they integrate with Logic natively using an existing or accompanying driver. To install a new device, go to Logic Pro > Preferences > Control Surfaces > Setup .... You'll face the Setup pane which has three menus at the top: Edit, New and View. Select New > Install from the menus at the top which will open another window listing all the Logic-compatible control surfaces. You are encouraged to "Add", or you can scan for attached devices.

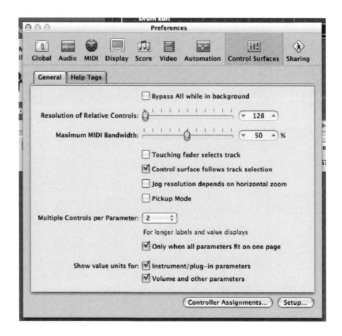

Once you have added your controller, close this window and return to the controller setup page. Here it is possible to select the controller from the dia-gram on the right-hand side and then edit any information about the device in the inspector-like space on the left-hand side. Within this space you're able to manage many aspects and preferences, such as what MIDI ports it is connected to, and what feature the fader controls within Logic.

It is possible to add more than one model of control surface. As such Logic makes it possible to edit what the controls actually operate in Logic within the setup page. In the example of more than one controller, duplicated buttons such as play and stop could be reassigned on one unit, whilst the faders could naturally lead on, say from Channel 8 on one device to Channel 9 on the next control surface. To get two control surfaces working in tandem, ensure in the Control Surface Setup window that the controller pictures form a horizontal row. However, should you wish for the controllers to act independently, then place the icons in a vertical column.

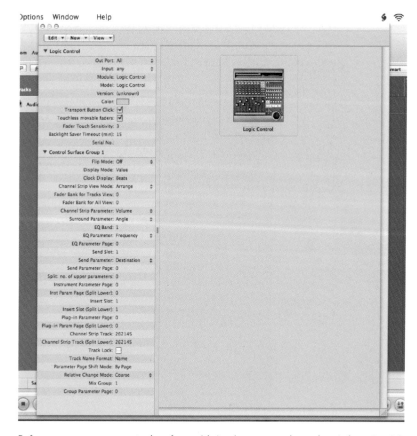

Before you can use a control surface with Logic, you need to select it from Logic's Setup page which can be found at Preference > Control Surfaces > Setup. With a controller chosen, it is possible to select its image and gain access to many editable features.

The way in which the Control Surface reacts and interfaces with Logic can be comprehensively edited using the Controller Assignments pane (Logic Pro > Preferences > Control Surfaces > Controller Assignments .... or simply press Apple + K). There are two views here, the first of which is Easy View. In this mode, each command can be seen one at a time and it is simple to reassign a control to another duty. To do this simply press on the part of Logic you wish to control. Next, press Learn Mode to move a control or touch a button on the controller surface to map it accordingly.

To edit these modes choose Expert View; the area to the left of the pane describes the zone of the surface, in this example the Mackie Control Universal. The control surface itself is split into zones such as VPots, Tracks (faders), Global Views, etc. The next table refers to the Modes in which each zone can operate and by selecting one of the modes, the Control/Parameter list will change to show you what each control does.

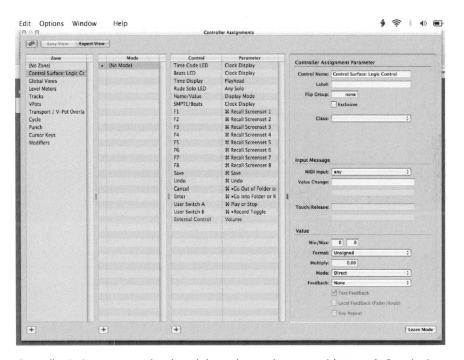

Controller Assignments can be altered dependent on how you wish to work. Reassigning controllers can offer a great deal of flexibility to tailor your control surface to the way in which you wish to work.

The Controller Assignments editor is incredibly powerful as any feature of the control surface can be remapped to another to allow you perfect control over a particular session or mix. In the example below, the selection list can be seen on the right-hand side offering a large number of choices for remapping an EQ control.

3.8 Digidesign hardware and integration with Logic

Digidesign, the manufacturers of ProTools, produce a number of audio interfaces which roughly fall into two categories. The first of these is their LE range, which is a host-based system, meaning that the DSP is handled within the computer's processor. For professional applications, Digidesign have pioneered the Time Division Multiplexing (TDM) system which makes use of DSP power held on their external HD cards, leaving the host CPU to get on with the work at hand.

Both of these types of system can be used with Logic as an audio interface using Digidesign's Digi CoreAudio Manager. However, further integration is possible with Digidesign's HD systems using ESB TDM. The Digi CoreAudio Manager effectively bridges between the hardware and OS X's Core Audio Services, allowing you to control the hardware as though you were in ProTools Hardware Setup page. When using the Digi CoreAudio Manager with Logic, it is worth noting that Logic and the MBox may not immediately talk to each other. Opening up Digi CoreAudio Manager before launching Logic should ensure communication.

Digidesign's HD range requires you to connect the external audio interfaces to their HD, PCI-based cards placed within your desktop computer. It is quite common to find both Logic and ProTools co-existing on the same computer, as they are used for differing applications at separate times, but can co-exist using the formerly named Emagic System Bridge, now simply "ESB TDM."

Logic does not automatically adopt such "external" technology. However, it is possible to access the Digidesign's Converters as though it were a host system. The HD Card's DSP capabilities are made available to the Logic system using ESB TDM using Digidesign Audio Engine (DAE) and direct TDM (DTDM). DAE is Digidesign-speak for the processing software that interfaces with the DSP chips inside the HD card. DTDM allows you to route some of your signals through to the TDM Mixer and process through the DAE, or TDM plug-ins. These signals can then be rerouted back to Logic to appear on some auxiliary channel strips.

To setup Digidesign hardware with Logic, go to Logic Pro > Preferences > Audio > Devices > DAE. Simply select the Enabled checkbox and you'll be asked to reboot to ensure that the hardware is all present and correct. This will allow you to connect to the Digidesign HD hardware that is attached. You should also tell Logic how many HD cards and interfaces are attached by clicking on the ProTools HD Type list (PT HD Type). Within this list there are details of the ProTools track count determined by the hardware and the amount of DSP chips are included on the HD card.

A good place to start when working with Digidesign hardware is to open up the TDM Configuration template from the Template Chooser. This offers you a mixer and tracks which will work with a Digidesign setup and allow you to get started. By employing this template, you can soon see how the audio interfaces and works with the Digidesign Hardware.

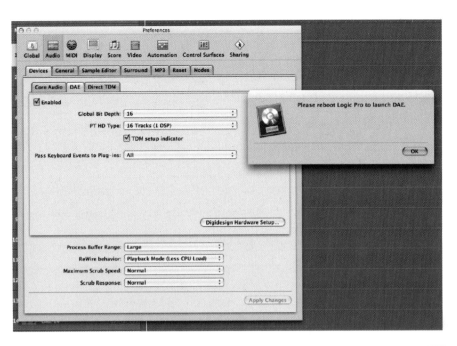

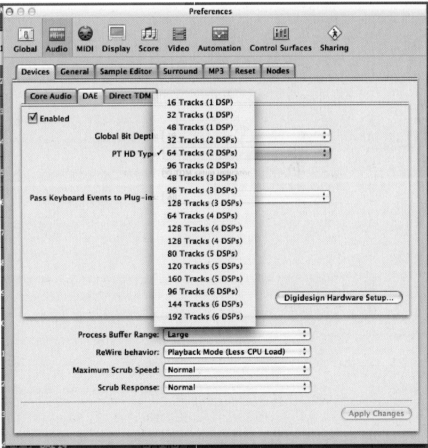

Setting up Digidesign hardware can be done within the audio preferences pane.

For more information on Digidesign HD integration with Logic, see the dedicated Logic Pro 8 TDM Guide which can be found in the Help menu in Logic.

When working with TDM hardware for the first time, it is beneficial to choose the TDM Configuration from the Template Chooser to see how Logic interacts.

3.9 Distributed audio processing and external DSP solutions

As with any project, there will come a time when the onboard processing power of the computer will start to strain under the pressure of plug-ins within your mix. As with Digidesign's HD system, there is an external solution to the problem for Mac users too. Distributed audio processing allows for an additional DSP to be made to the computer either by Ethernet, such as the Logic's Nodes' or by an external DSP solution connected via FireWire or PCI.

Using a system such as Solid State Logic's Duende or Universal Audio's UAD system (shown opposite), power can be unleashed from your computer to make for a faster and more powerful workflow. The audio processing is taken care of outside of the host computer using proprietary processing to each manufacturer. As such plug-ins can be purchased to add to the systems which might be more reliant on your DSP than you would desire for your computer.

Setting up an external processing device such as SSL's Duende is easy and is often a matter of simply connecting the FireWire and installing the accompanying driver. With OS X rebooted, Logic should now pick up that it has Duende installed. Logic may rescan its audio units and an extra line within the audio units menu when selecting plug-ins called Solid State Logic. Within here should be the features that are available on the external unit such as the SSL channel strip.

SSL's Duende and Universal Audio's UAD systems are excellent examples of distributed audio processing for Logic users. (Images courtesy of Solid State Engineering).

For more installation guidelines, please refer to the manufacturer's instructions and website.

Using multiple Macs

Distributed audio processing can be achieved by employing multiple Mac computers and connecting them by gigabit Ethernet. These are then called "nodes" in Logic speak and will allow for your processing tasks to be distributed across a unique network. Setting up nodes and multiple Macs is covered in Chapter 11. This is very flexible as it allows you to harness the power of your older Mac or desktop workhorse, but still allows you to take your laptop out on the road.

Knowledgebase 3

ReWire Connectivity

ReWire is a system developed by Propellerhead, the makers of Reason and ReCycle. ReWire cleverly allows internal connections to occur between audio software packages within one computer. It is therefore possible to have multiple MIDI connections internally between applications and receive the corresponding audio signals back in return. This is an incredibly powerful and flexible solution when getting the best from two applications. Setting up ReWire in Logic 8 is very easy. You must first ensure Logic Pro 8 is opened up before the ReWired application such as Reason or Ableton Live.

With Logic opened, launch the ReWired application, thus ensuring that Logic knows that this is to be the ReWired application and will therefore make the appropriate connections for you. To use your ReWired instruments, first create an External MIDI track by pressing Apple + Alt + N on the Arrange Area. Next, call up the Library Tab from the Media Icon on the top right hand side of Logic. Within the Library should be your opened ReWired application, in this instance let's say Reason. Clicking on the Reason folder in Logic's library should reveal all the open Reason instruments. Simply click on one to map this to the selected MIDI track.

As you select new instruments within your ReWired application, both the MIDI and audio connections should be automatically made to Logic. The audio connections will appear as Auxiliary tracks on your Logic Mixer and can be processed using Logic's plug-ins and blended into the mix as though they were another virtual instrument.

Walkthrough

Setting Up Aggregate Audio Devices

Step 1:

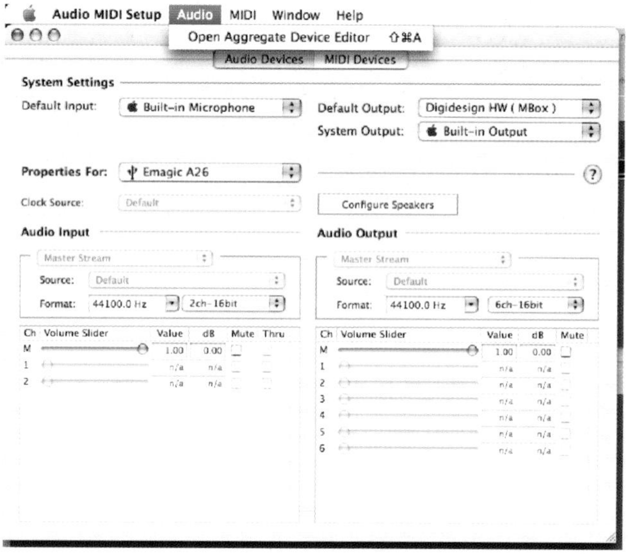

In situations where you want to increase your possible inputs and outputs, you may wish to create what is called an Aggregate Device. Your Mac will invisibly pair these devices together to create a new "aggregate" with their inputs and outputs added together. From the Finder, open the Utilities menu (Apple + Shift + U) or Go > Utilities. Next, open up the AudioMIDI Setup. The Open Aggregate Device Editor can be found in the Audio Menu or by pressing the Apple + Shift + A key command.

Step 2:

Starting at the large "+" sign near the left of the page, we can create a new Aggregate Device. This can be renamed at any time by clicking on the name. Below is what Apple refers to as the Structure of the Aggregate Device. In this table, you can indicate which devices will be included in the Aggregate, specify the master clock of all the devices and whether they should resample the signals to match the master clock source. This is due to the fact that in order for aggregate devices to act as one, their clocking needs to be matched. Therefore it is important to connect the digital out-of-the-master device to the digital in-of-the-slave.

Step 3:

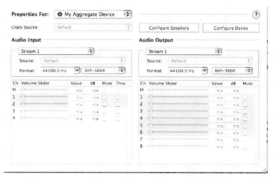

With the Aggregate Device now set up we need to select this from the inputs and outputs lists within the Audio MIDI Setup if it is required by other applications or by the operating system. It is important to select the Aggregate Device in the "Properties For:" menu and edit the inputs and outputs as required. If we now restart Logic and navigate to the Audio Preferences (Logic Pro > Preferences > Audio > Devices), we can now select the "new" device from the menu for use within Logic.

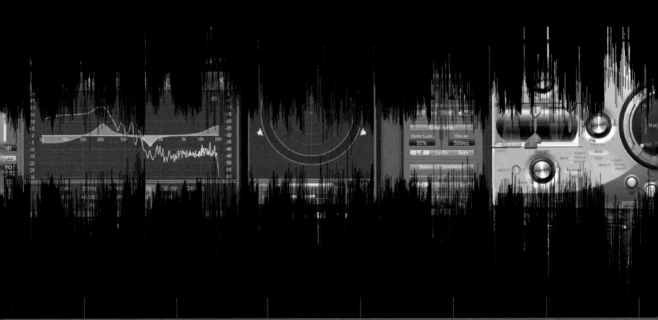

Starting a project

4.1 Introduction

Laying the successful foundations at the start of your project is vital to the long-term success and seamless workflow of realizing audio and music production work in Logic. Of course, not all projects start the same way, but we're going to begin by looking at the process of recording audio into Logic – in effect, using it as a modern day replacement to the multitrack tape machine. This is also an excellent way of covering some important essentials of working in Logic, like creating tracks, using the transport controls, the bar ruler, and basic monitoring. We'll also start to explore the various aspects of data management involved in working with Logic – more specifically, how and where you should store your project data, and what a Logic project is actually comprised of.

If you're short of inspiration, or just needing something to kick-start your creative process, we'll also take a look at importing and working with Apple Loops. Apple Loops – contained within the multitude of Jam Packs installed with Logic Studio – are an instant resource of loops, instrument, and vocal passages, all copyright-free and ready to be used in your compositions.

Moving beyond the basics of track laying, we'll also take a look at the various ways you can make a recording session run even more smoothly – providing a dedicated headphone mix for musicians, for example, or using plug-in effects to better understand how your instrument might eventually appear in the mix. All of these factors will create a much more comfortable recording experience for you and your musicians, allowing you the musicianship, engineering and performance qualities to become more focused.

Of course, this is only the starting point to your creative work in Logic, as we'll also be looking at how to edit this newly recorded material in Chapter 5, and then on to integrating the range of virtual instruments and MIDI sources in Chapter 6.

4.2 Assets and projects

The basic currency of Logic is a Project, which will contain all the relevant assets for the song you're creating. In respect to Logic, the term "assets" corresponds to the collection of separate files and data that a track can be comprised of. For example, as well as the song file, which contains all the important arrangement information, you could easily assemble any amount of audio files, sample data (for the EXS24 sampler), and movie files, all of which will reside in the main project folder. By organizing your data in this way, both you and Logic can keep better track of your creative process, making the process of backing up, for example, or moving to a different Logic 8 setups, considerably easier.

The starting point of your creative process is to create a new project (File > New). Logic will then present you with a series of possible starting points, including some interesting options for composing or producing music in Logic, either based on genre or production objectives – like surround mixing, for example, or mastering. If any of these templates meet your particular needs, then it might be appropriate to select them, otherwise select the Empty Project as a suitable initialized starting point for production. Note that you can also automatically select this option by holding down Alt as you select New from the File menu.

Creating a new project (File > New) will present a number of starting templates. Select *Empty Project* if you want a simple initialized starting point.

Before you go any further, though, it's worth saving your project, if only to allot a particular location for your project folder. If Logic hasn't already opened

Knowledgebase 1

What are assets?

Assets are Logic's collective term for the various data files that can be associated within a project. Of course, the project file – or song – is only the beginning of the data you'll amass over time, with the full set of assets including Audio Files, Movie Files, EXS24 Sampler Instruments, Ultrabeat Samples, and Space Designer's Impulse Responses. Ultimately, if you don't have some organized strategy for looking after these files, you'll soon find that a project becomes difficult to manage, especially when it comes to archiving a project or moving it to another system.

By keeping the files in a single folder, Logic offers at least one way of organizing this data, although if you've got an alternative system, you can always decide to manage this yourself. Theoretically, this can prevent certain unnecessary file duplication, which can easily become an issue with large EXS24-based projects. To change the current project's Asset settings at any time, select File > Project Settings > Assets, where you can switch in and out different parts of the Asset management accordingly. By default, Logic concentrates on audio files, including automatic sample rate conversion of files imported that don't match the current project settings.

Managing the Assets related to your project will allow you to create better project archives as well as move between different Logic setups in a more effective way.

a Save As dialogue, you'll need to do so (File > Save As ...). Besides specifying the name of your project, you can also instruct Logic to automatically save the relevant assets to your project folder, under the Include Assets option. It's also well worth checking the advanced options, as this will allow you to differentiate between whether Logic simply collates your respective audio files, or added extras like EXS Sample Data, Instrument files, Impulse Response, Movies, and so on. Ultimately, if you're unsure, there's no harm in selecting all the options, but this will use a correspondingly larger amount of drive space.

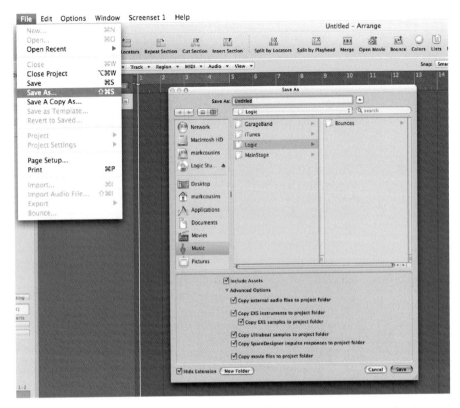

The Save As dialogue allows you to create a root folder for your project, as well as to check what Assets will be saved with it.

While all of this initial project management can initially appear time consuming (especially if you just want to get on with making music!), it's well worth pointing out that good data management is a vital part of music making on a computer. Spending time to ensure you know where your data is being stored will reap plenty of dividends in the long run, making it far easier to recall projects several months down the line, or indeed, work more effectively between different computers.

4.3 Working with tracks

Assuming you've selected an Empty Template, the next task is to create a series of tracks to record your audio and MIDI information into Logic. For example, if you're recording a band using a number of microphones, you'll need to create an accompanying collection of audio tracks representing each instrument or mic you're using. If you intend to create a track largely using virtual instruments, with a few vocal overdubs, you'll want to create a 16 or so instrument tracks combined with 1 or 2 audio tracks. Of course, you can always add to, remove, or re-order tracks from your project at any point in the production process, so don't feel obliged to pre-plan your complete track list at this point.

To create a track, click on the small + sign on the top of the track list, or, if no tracks are present, Logic will ask you what tracks you initially want to create. The dialogue box that accompanies track creation allows you to specify both the exact type of track that is being created, as well as several options for creating multiple tracks, output and input assignments, and so on.

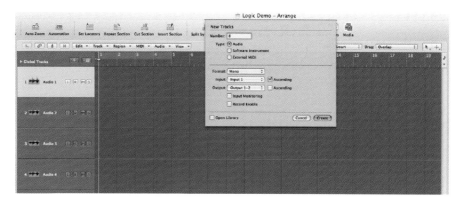

Create a number of empty tracks to correspond with the instrumentation you want to record. The New Tracks feature includes several ways of speeding this process up.

Audio

An audio track will allow you to record an audio signal being fed from a corresponding input on your audio interface. This is also the primary focus in this chapter.

Software instrument

Use a software instrument track if you want to record music using any of Logic's integrated virtual instruments (EXS24, EVP88, ES2, and so on), or indeed, any third-party Audio Unit instruments. We'll be taking a closer look at the process of working with virtual instruments in Chapter 6.

External MIDI

External MIDI tracks correspond with MIDI hardware like synthesizers and samplers that you have externally connected to Logic (usually via a USB interface). Again, we're going to spend more time looking at this functionality in Section 6.5.

As you're creating your first set of audio tracks, you'll notice a couple of important options to speed up your workflow. For example, if you intend to record from the first 16 Inputs of your audio interface, you might wish to increase the Number (of track) parameter to 16, and click on the Ascending Input option – that way, each subsequent track will take a different input number. On the whole, most users tend to monitor their session from a single Output, so you'll probably want to leave the Output setting out of its Ascending option. Whether its Format is mono or stereo will depend on the type of signal input being fed to your audio interface – a pair of drum kit overheads, for example, will be stereo, while a vocal (recorded with a single mix) would be mono.

With your basic tracks created, you'll want to also create some initial track names. Not only will this make the navigation of Logic's arrangement window and mixer much easier, but you'll also stand much more chance of keeping track of the various project audio files with names the "kick drum" rather than "audio 01." To name a track, simply double click on the existing track name as part of the track list. Once named, any subsequent recordings on that track will adopt the name accordingly.

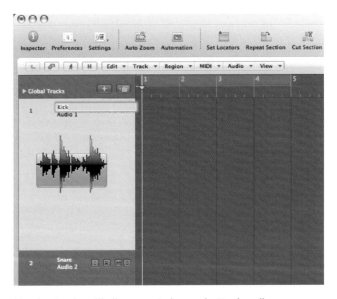

Naming tracks will allow you to keep a better handle on your logic arrangement, as well as being used to automatically name the audio file and region you create.

4.4 An introduction to the Audio Mixer

Although we don't need to go too deeply into mixing at this point, it is well worth introducing the concept of the audio mixer, mainly as an alterative way of looking at the tracks you've created for your project. While the track lanes of the arrange area work horizontally, the mixer area shows us our current project in the form of a traditional vertical mixer, which each track represented as a different channel strip. You can open the mixer at any point using either the mixer tab at the bottom of Logic's arrangement window or from the keyboard shortcut X.

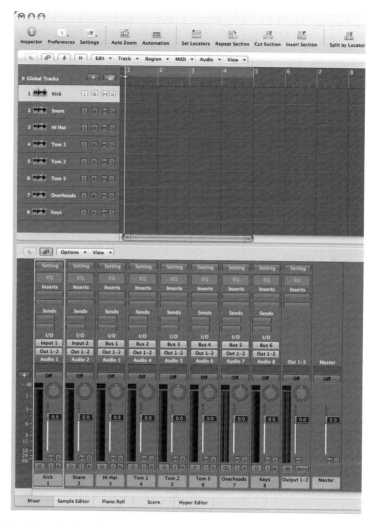

The mixer window (accessible from the mixer tab, or by pressing X) displays an alternative view of your newly created tracks.

Common to both the mixer and the arrange area's track list are a series of function switches, each of which govern how the particular track or channel is working at any given point in time.

Input Monitor (I)

Input Monitoring is a simple one-click means of directly monitoring, or listening to, the current selected input for that given track or channel. At first, this might be a good initial way of establishing what's coming into your inputs, and, using the meters on the mixer strip, setting an initial gain on your microphone and line inputs. Note that in most cases, this will be done directly from the interface itself (or the interface's control software, if you're using something like the digitally controlled Apogee Ensemble) rather than adjusting the fader's position. In this case, the fader is simply setting the monitor level, rather than any form of recording level as it might do on a conventional console.

Should you wish to change a track's particular input, this can be done from the mixer, clicking on the small I/O box above the fader. The list presented should tally with the inputs available on your audio interface. For example, a typically USB audio interface will present 2 available inputs, whereas a FireWire interface could offer 10 or more possible inputs. Clicking on the small circle button on the bottom of the fader will also change track between mono and stereo operation, with a corresponding change in the input selection – input 1, for example, would become input 1–2.

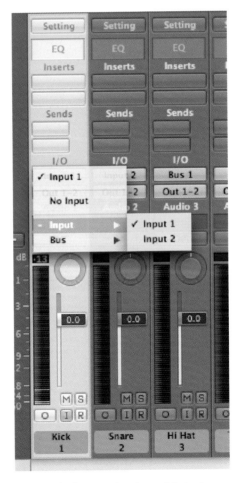

Use the I/O box on the channel fader to change a track's input status, with the number of inputs corresponding to the type of interface you're using with Logic.

Record Arm (R)

Record Arming a track or channel indicates to Logic that you're about to make a recording. Again, you'll be able to see your input levels accordingly, and hear the signal routed through to your main outputs. Once you press record on the transport, though, Logic will actively be recording the input as a new audio file (more on this in a moment!).

Mute (M)

As you'd expect, mute silences the track or channel in question. What can be confusing, though, is how this differs between the arrangement area and the mixer. As the controls are independent (unlike the Record Arm or Input Monitor options), you won't find changes on one reflected on the other. Instead use the mute to silence the track or channel. For example, you might run several track lanes for the same instrument (like different takes of a vocal) where mute would silence one of those. Muting the channel, however, would silence any signal being directed to that strip.

Solo (S)

Solo will play the chosen track or channel in isolation, and, like the mute functional, works independently so that you can either solo an individual track lane or a number of track lanes being fed to the same channel strip.

Working between the audio mixer and the arrange area is an important part of the day-to-day usage of Logic. For example, note how a selection in either window is mirrored in the other – selecting the vocal track on the arrange area, for example, will also bring it up on the mixer.

4.5 Using the transport and timeline

As with any multitrack recorder, the transport bar – found towards the bottom of the arrange window – is what provides the master control for recording, playing back and looping in Logic. For example, having created your tracks and record-armed them, pressing Record on the transport will tell Logic to engage its playback (of any existing material) and record each track as an individual audio file. The rest of the controls (play, pause, rewind, fast-forward and so on) need little introduction, although it's well worth noting how the transport controls are also influenced by settings in the arrange area's Bar ruler.

On the right-hand side of the transport, you'll find a number of different Modes and Function buttons – like a low-latency monitoring mode, Solo, and of course, the metronome. We'll be covering these in more detail as we move through the recording functions in Logic.

The transport bar also includes a number of important mode and function buttons – including the Metronome and Autopunch – that are vital to the process of recording in Logic.

Bar Ruler

The Bar Ruler – along the top of the arrange area – provides the positional reference, or timeline, for our project, with the Playhead indicating the current song position. As you'd expect, there are a number of additional functions of the bar ruler, allowing you to specify parts of the recording to be dropped in, for example, or areas to be looped around. Crucial to the functionality are the left- and right-hand locators, displayed numerically on the transport, or the lighter grey shading in the bar ruler.

You can adjust the left and right locators in a number of ways, although probably the quickest and most logical solution is to drag from left to right in the top half of the timeline. As you drag your selection, you'll automatically place Logic into cycle mode – as illustrated by the both green shading in the bar ruler, and the fact that the cycle mode indicator (found in the bottom right-hand side on the transport) is active. In cycle mode, you can repeat over a section of the music, and if you're recording, lay down a number of different passes or takes of the same musical line to be edited later on.

Here are two left and right markers defining the length of our cycle. Use the cycle feature either to loop round a portion of the song or perform a series of different takes.

To disable cycle mode, you can click on the green region in the bar ruler, or alternatively deactivate cycle mode on the transport. Even when it's deactivated, though, Logic will retain the left- and right-hand locator settings allowing you to return to the cycle at any point. Further to this, you could also define a number of different cycles using the Marker feature (like Verse, Chorus, and Middle Eight), which we'll cover in more detail in Chapter 10.

Tip

As well as using the on-screen controls, you can also use Apple Remote Control to operate the transport of Logic. Using your infrared control you can remotely start and stop the transport, move backwards and forwards bar by bar, and toggle through the various track lanes. To record, simply press the Menu button – a great solution for anybody recording themselves away from their computer!

4.6 Your first recording

Assuming you've created and record-enabled the appropriate amount of audio tracks, you should be ready to make your first recording. Recording will begin in one of two places – either at the current position of the playhead or, if you've set up cycle playback, at the point the cycle begins. To provide you with a small amount of pre-roll, Logic will actually start playback a bar earlier than current record point, allowing you to get some indication of the track's speed, as well as existing elements of the performance that might lead up to it.

When you've finished your first pass, take the track out of record arm mode. If you've got multiple tracks record-enabled, this can be done by Alt-clicking on any one of the active record-enabled tracks. Playing the project back will now allow you to audition the takes, spotting any deficiencies or areas you might want to improve.

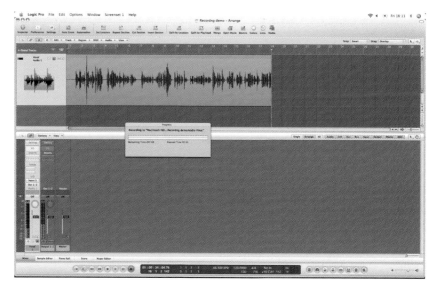

Recording the first pass into Logic.

Knowledgebase 2

More about the Metronome

In certain situations, you might find it beneficial to adapt some of the qualities of the metronome – stopping the metronome being sent over MIDI, for example, or changing the preset Klopfgeist sound to something of your own liking. The Metronome settings can be accessed via the Project's settings (File > Project Setting > Metronome), or directly, by Ctrl-clicking on the Metronome icon on the transport bar. The settings allow you to switch the Metronome in and out of MIDI playback, as well as being able to disable the software click and adjust some basic parameters like tonality and volume.

Delving deeper in to the audio mixer environment (which we'll cover in more detail in Chapter 11, but for now access via Window > Environment) you can also change the default Klopfgeist instrument assigned to the Metronome. Klopfgeist is a simple instrument, specifically designed to produce distinctive click sounds that can be audible over a mix, yet not so loud that they would bleed out excessively over headphones. If you've got the mixer environment open, you should see the Klopfgeist instrument assigned to the Click object. By changing this to the EXS24 sampler, for example, you could replace the standard Klopfgeist click with a rimshot sample from one of the factory kits.

Adapting the metronome settings can be achieved via the menu option File > Project Setting > Metronome.

Difference between mix levels and monitoring levels while recording

One aspect that you might have noticed is that Logic effectively stores two fader positions for each channel – the level for recording and the level for playback. For example, as you're recording the part, you might find it beneficial to increase the level of track for monitoring purposes, that way you can hear the instrument being recorded more clearly. In playback, however, you might want to reduce the level so as to hear the part better in the context of the mix as a whole. Move back to recording again, though, and the original record monitoring levels will be restored.

4.7 Overdubs and punching-in and -out

Although we'd all like to think we could record perfect "one-take wonders" every time we enter the studio, there's always an inevitable amount of remedial recording that needs to take place – either with fluffed notes or whole sections that need to be replaced. Logic, of course, provides a number of solutions for this happening, all of which can be further refined when it comes to editing your project (as we'll see in Chapter 5). Unlike tape, it's worth remembering that any punch-ins or overdubs are "non-destructive" – in other words, the original takes remain untouched on the hard disk, with Logic simply changing or augmenting the files that it decides to playback for any given track.

Quickly punching-in material

One solution to repairing your recording could be simply to record over a section of the track in question. To do this, all you'll need to do is place

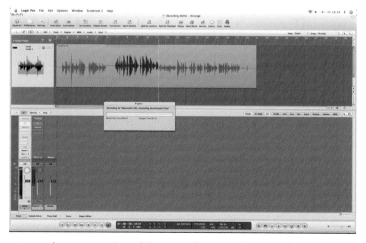

If you want to replace any section of the recording, simply locate the transport at the point where you want to drop in, and press record. Logic will provide a bar of pre-roll and then begin recording your new material.

the track or tracks you want to replace into record-ready mode, locate the playhead at the punch-in point, and press record. Logic will then give you the usual bar pre-roll and then initiate recording, effectively "replacing" any previous recording you have made at that point (although remember, this is non-destructive). When you press stop, you'll effectively punch-out, leaving subsequent audio recordings untouched.

Recording in cycle mode

If you've got a specific section of the song you want to work on, it might be appropriate to record in cycle mode. In cycle mode, Logic will continually loop the portion of the song, recording each pass that you make. When you've stopped the transport, the last take will be used for playback, although all the various takes will have been recorded.

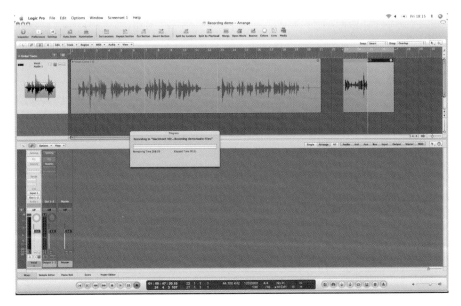

Using the cycle function is a useful way of loop-recording a number of bars until the part is performed correctly.

Autopunch

If you're clear about a specific part of the track you want to replace, consider using the Autopunch feature. Autopunch can be activated via the transport bar, with an additional set of locators viewable in the middle of bar ruler. You can change the position and length of the Autopunch locators in the same way as the cycle locators, ideally setting the locator points to precisely match the portion of the track you want to replace. With the track record-enabled, and record engaged on the transport, Logic will playback the song and only record for the precise duration of the Autopunch locators.

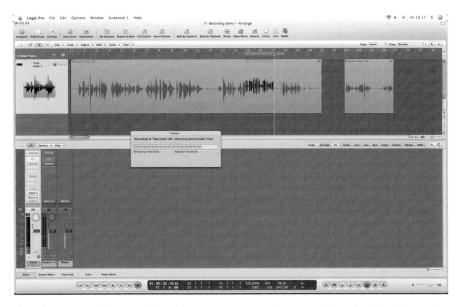

If you have a specific part of the take you want to replace, consider using the Autopunch feature for an accurate, automatic drop-in and drop-out.

Introducing Take Folders

Whenever you overdub onto an existing recording on an audio track, or use the Autopunch feature, you create what Logic refers to as a Take Folder. If you look closely at a take folder, it looks slightly different to a conventional audio recording in Logic, with two triangles in the top left- and right-hand corners of the region. Technically, a take folder stores all your various takes in an organized way, with features both to switch between takes, or make a quick edit between the different takes.

To understand how we can carry out further work with the take folders, take a look at Section 5.4 – Quick swipe comping.

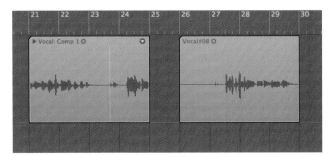

If you record over an existing region, Logic will create what is known as a Take Folder. A Take Folder will allow to reposition and refine edit points later on in the production process.

4.8 Creating further tracks and track sorting

Although we should have created the majority of tracks required for our session at the start of the project, it is highly likely that you'll need to add further tracks into your session. Let's take a look at some of the motivating factors and the quick steps to create and organize these tracks.

Creating duplicate tracks

Use the menu function Track > New with Same Channel Strip/Instrument (or Alt + Apple + S) to create a parallel track with the same instrument or channel assignment. This is a useful tool if you want to create two alternate "takes" of the track – so, rather than packing different takes in a take folder, you end up with two discrete takes assigned to the same channel. Remember though, that only one audio recording can take priority at any given point in time, so you'll need to be clear what track you're listening to by muting the unwanted takes accordingly.

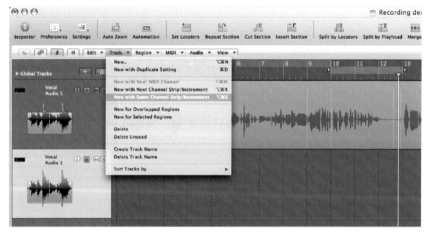

Creating a duplicate track is useful way of negating the Take Folder system, allowing you to clearly visualize each take as it's being recorded.

Creating tracks with the next instrument

Creating tracks with the next instrument or channel strip (Track > New with Next Channel Strip/Instrument) is a great way of augmenting your existing project, adding further track lanes for each new instrument you might want to add into the mix. Using the keyboard shortcut, Alt + Apple + X, allows you to do this quickly and efficiently without interrupting your creative workflow.

Creating tracks with duplicate settings

If you've already set up some particular monitoring EQ and compression effects, for example, you might want to create new tracks with duplicate plug-in settings (Track > New with Duplicate Setting). The newly created track will have it own unique channel assignment, but a copy of any of the plug-in settings as well as any fader or pan positions you've established.

Sorting and organizing tracks

As the project grows, it might become beneficial to organize your track list into an order that better corresponds to the arrangement of instrumentation, as apposed to the order you recorded it in. For example, maybe you want to sort all the vocal tracks together with the lead vocal first, or separate the keyboard from the guitars. Arranging the track order is as simple as clicking on a track lane and shuffling it up or down the track list accordingly. Should the track list become increasingly confusing, you might also want to return it to something closer to its original arrangement. In that case, go to Track > Sort Tracks by > Audio Channel, for example, and have the tracks automatically re-ordered by their channel assignment.

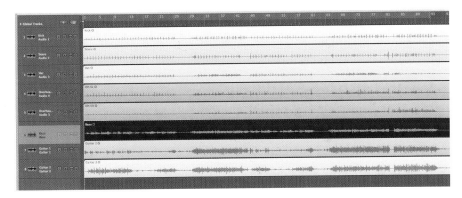

Try sorting tracks to make their organization clearer.

4.9 The Audio Bin and importing

While the arrangement area provides an important positional overview of the audio files we recorded as part of our project, it's not the only way we can manage and organize the various files we've created. The Audio Bin presents the complete list of the audio files included in your session, as well as offering a means of importing new audio files (possibly from another project, or a drum loop from an audio CD). To open the Audio Bin select View > Media, or open the Audio Bin using the keyboard shortcut B.

Scrolling down the Bin you should be able to see all the files that are contained within your project, even ones that might not be currently active in your arrange area. Alongside the file name, you can also get some basic information on the audio file, including its sample rate, bit depth and file size. Another important piece of information is the file's location, which is displayed in the top line whenever a particular file is selected. If, as part of your asset management, you specified for all files (recorded or imported) to be stored in your project folder, then all these locations should be the same. If the project's data management isn't quite so organized, this might be an opportunity to find out which files are located outside your main audio files folder.

The Audio Bin provides an effective overview of all the audio files used within a project.

Importing other audio files and CDs

As well as viewing existing audio files, you can also use the Audio Bin to import new audio material into your Logic project. Logic supports a number of different audio file formats – including WAV, AIFF, SDII, CAF, MP3, AAC, SMF and REX2 – all of which can happily reside together in the same project. To import a file into your project, go to Audio File > Add Audio File and browse your drives accordingly. Again, if you've got your asset setting established correctly, the files will be copied into your project's audio files folder, to keep all the media content in a centralized, project-specific location.

To finish the importing process, you will, of course, need to drag the audio file into your project's arrangement. The precise method you choose to do this will have an effect on how the resultant file is played back. If you drag the file onto an existing audio track, the new audio file will be played back with the settings on that current channel strip. However, if you drag the new region to a position in the arrange area without a track assignment (this would be at the bottom of the track list), Logic will create a new track specific to the audio file in question.

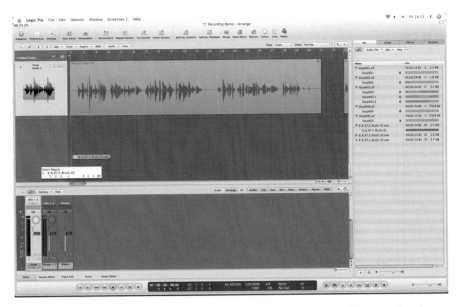

Use the Audio Bin to import additional audio files into your project. These can then be dragged into your arrangement with a new track created accordingly.

4.10 Working with Apple Loops

Apple Loops provide a quick-and-easy source of inspiration for music and audio production, and can be an effective way of kick-starting a project. Included with the full install of Logic Studio are a number of Jam Packs providing everything from Urban drum loops to Foley effects. The important point to note about the Apple Loops format is that the tempo and/or the key of the loop will automatically conform to your project's current settings, allowing you to quickly audition a loop's relative merits without having to time-stretch the file to fit. Equally, if the project's tempo changes at a later point (even midway through the song!), the Apple Loop will follow these changes accordingly.

You can access the Apple Loop browser under the Loops tab of the media area. The Apple Loops browser is an extremely effective way of exploring the content installed with Logic Studio, allowing you to specify various search tags to sift through the huge amount of possible sound content. The browser itself can be organized into the different viewing styles – a Column view, Music view, and Sound Effects view. Arguably the most intuitive of these is the Music view (indicated by the small musical notes), which organizes the various loops by genre (Jazz and Country, for example), instrument (Vibes, Strings, and so on) and sound-type "categorization" (Melodic, Distorted, Clean, and so on). By clicking on combinations of these tabs you should be able to arrive at a loop, or collection of loops, that best suits your needs. Pressing the Reset tab (top left) will remove any existing selections, allowing you to perform a new search.

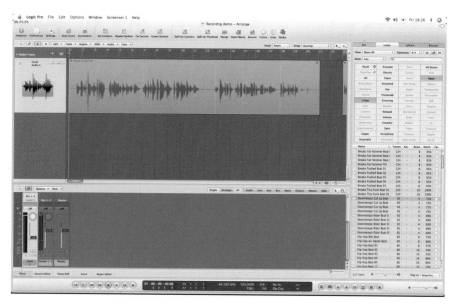

The Apple Loop browser allows you to quickly locate different loops based on their musical qualities.

Knowledgebase 3

Importing REX2 files

The REX file format was originally developed by Propellerhead as a cross-platform solution for transporting time-sliced audio files created by their popular ReCycle application. A number of third-party loop libraries make use of this file format, so it's useful to know how Logic handles these files.

One option is to load the REX file directly into the arrange area itself, with the slices left as a series of different regions locked to your particular project tempo. To import the loop, go to the Browser and drag the REX file into its first position in the arrange area. Logic will then ask you to confirm the REX file import, possibly suggesting a series of crossfades to best match the original loop tempo with the tempo of your session. If you confirm, this will create a folder in the arrangement containing all the accompanying REX slices.

As an alternative to importing the slices, you can also choose to render the REX file as a complete file or an Apple Loop by selecting the appropriate option from the pull-down menu. Arguably, a rendered audio file is easier to manage than a collection of slices, although it won't be able to follow changes in the project's tempo. As an Apple Loop, though, the REX file will retain its elasticity, allowing it to match the project's tempo at any point in the creative process.

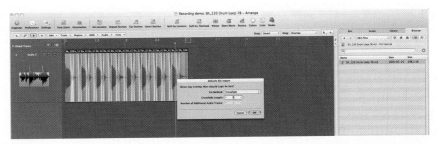

REX files can be directly imported into your Logic arrange, with a number of different options to conform the loops to your project's tempo setting.

The alternative browsing methods allow you to explore the Apple Loops sound content in different ways that might better suit your methods of production. For example, the Sound Effects view (found under the FX tab) uses a variation on the music view, with a replacement set of categories that better relate to the sound to picture work. The Column view, on the other hand, sifts through the sound content in a way similar to the Finder in its column mode.

The Sound Effects view provides an alternative way of navigating and viewing your Apple Loops sound content.

To hear an Apple Loop simply click on its file name, with Logic playing the file back at the same tempo and key as the song you're working with. You can also use the small volume slider and the bottom of the loops browser to set the volume for auditioning the loop. It's also worth checking the percentage match, as this will have some effect on the relative quality of the loop in the final arrangement, with loops closer to the project's tempo requiring less time compression or expansion to get them sitting correctly.

Audio files versus MIDI files and instruments

The Apple Loop files format actually contains two variations – a pure audio file that offers a recording of the loop in question and a MIDI version (also known as an Apple Loop+) that combines the raw note data with an accompanying software instrument and effects. You can see what type your Apple Loop is by looking at the small icon to left-hand side of the file. A blue waveform icon indicates that the Apple Loop is only available as an audio file, while a green music note indicates the Loop is available as raw musical data (in other words, a MIDI version) and as an audio file.

In the long run, a MIDI-based version will provide much greater flexibility than an audio-based Apple Loop, allowing you to precisely edit the musical data, instrument settings and effects. However, these versions can, at times, make slightly larger demands on your CPU usage.

Here are the two Apple Loop file formats and how they differ when imported into your session. A standard Apple Loop (top) is audio-only, but the enhanced version (bottom) contains the specific MIDI and instrument data used to create the loop.

Importing Apple Loops into your session

To use either of the two Apple Loop formats, all you need to do is simply drag the file into your arrangement. One point to be clear on, though, is the type of track you import it to and how this has an effect on the end form of Apple Loop imported. For example, if you drag the Apple Loop onto an empty track lane with no track assignment, Logic will automatically create the appropriate type of track (Audio or Instrument) based on the particular Apple Loop in question. By dragging the Loop to a specific type of track, though, the Apple Loop will be conformed to the track type – so, for example, MIDI-based Loop would be placed as its audio version on a corresponding audio track. The only exception, of course, is that audio-only loops can't be converted into MIDI regions.

Once you get more proficient in Logic, you might also want to consider creating your own Apple Loops files, which allow you to access the same "elastic time" properties with your own audio material. For more information on this, see Section 5.13 – Making Apple Loops, in Chapter 5.

4.11 Improving what the artist hears – headphone mixes

Although we've looked the process of basic track laying in Logic from the engineer's perspective, it's well worth considering what exactly the artist hears and performs to as the material is being recorded – in respect to either their individual headphone mix or the addition of plug-in effects – like compression, distortion or reverb – that might be added during the process of recording. However, it's worth pointing out that these subsequent techniques will push your Logic skills much harder than the first part of this chapter, particularly with respect to the use of the mixer.

First off, let's consider the two types of mixes heard in a typical tracking session in a professional studio.

Monitor mix

The monitor mix is simply the mix that is heard in the control room, and is usually based on a reasonable and considered balance of the material that has been recorded. In some cases, the producer or engineer might want to hone in on a specific part (usually whatever is being recorded at that point) with the rest of the instrumentation significantly quieter. Ultimately, a good monitor mix is the only way you can ensure you're making an effective recording.

Cue Mix

The cue mix (also known as a talent mix or headphone mix) is the balance of material that relates specifically to the performer and what they hear through their headphones.

Of course, a pragmatic solution to recording sometimes involves the monitor mix and the cue mix effectively being the same thing, especially if you're recording yourself! If you're using a USB audio interface, it might also be the case that the interface features some direct provision to control the mix between the input signals and the DAW's mix so as to combat any latency issues (for more information on Latency see Chapter 3), again resulting in a combination of both Monitor and Cue Mix. However, if your audio hardware supports it, and you need a more professional amount of monitoring control, it's far better to create a separate monitor mix and cue mix directly from Logic.

Creating a unique headphone mix in Logic involves a more detailed use and exploration of the mixer page, so as to effectively create two mixes at the same time. On each of the channel faders in your project, you should find a series of two empty boxes under the Send section. As we'll see later on the mixing chapter, these bus sends, as they are called, have several different applications, but for now, we're going to concentrate on their role in the creation of a cue mix. To create a bus send for a particular channel strip, click in the small grey Sends box and insert a send to Bus 1. Note that if you have a large number of tracks, you can also select multiple channels, insatiate the Bus on one of the channels, and have them all duplicate the same action.

Use the Sends box to create a bus sends for each individual channel in your mix. The bus sends are the controls that will be used to create your headphone mix.

The next important thing to do is to set the pre/post assignment of the bus sends. Although this sounds complicated, all a post/pre assignment means is whether the bus send is patched before or after the channel fader. Thinking through logically, it makes most sense the cue mix is created pre fade – so that any movement of the channel faders won't immediately change the relative mix for the headphones. In other words, if you want to turn down the vocal in your monitor mix, the vocalist can still hear himself or herself loudly and clearly. The pre/post assignment can be easily switched by clicking on the bus send and changing the mode from post (the default setting) to pre. If this is done correctly, the colouring of the send should change from blue to green.

Switching to pre fade will allow the mix to be completely independent of Logic's channel faders. The colour of the bus sends indicates the pre/post status, with green being pre fade.

In creating the bus send, Logic will have also created a corresponding Aux master object, which will act as the master fader for you cue mix. To make things clearer, you could double click on the channel's strip name towards the bottom on the mixer and change this to something like *headphones* or *cue mix*. To build up the headphone mix, solo off this channel and then blend the appropriate instrument using the bus sends on the individual channels. As a good tip, maybe start by placing the loudest possible instrument in at around 0 dB (you can do this by Alt-clicking the corresponding bus send pot), and then sitting the rest of the instrumentation around this.

To send the headphone mix to the specific physical interface output, change the output status of the aux master accordingly.

4.12 Monitoring through effects

In some cases, performing a take can sometimes be tricky without an idea of how it will sound in the final mix. Although it is often the intention of the engineer to record "dry" sounds devoid of artefacts like compression or reverb (which will instead be added in a controlled way in the mix), the same cannot be said of what the artists themselves prefer to hear. Far from being fussy, effects like reverb can actively help a vocalist pitch themselves, equally the drive and distortion of a guitar sound is such an integral part of how a guitar is "played."

Effects plug-ins can be instantiated using the small grey Inserts boxes in the mixer's various channel strips. Clicking on box will bring up a list of Logic's own internal plug-ins (organized into groups – like delay, distortion, filter, and so on) as well as any third-party Audio Unit plug-ins that you have installed on your system (organized by manufacturer). You can also activate EQ at any point by double-clicking on the corresponding EQ box just above the inserts. An important point to note is the order of the plug-ins – reflected in the progression from top to bottom on the inserts – which will dictate the order in which an instrument is processed. If you're a guitarist, for example, you might want to place certain effects before and/or after your virtual amplifier (in this case, Guitar Amp Pro). Alternatively, a vocal track tends to have EQ and compression before reverb, so that the reverb isn't equalized and compressed.

We'll be taking more of a look at the art of setting up effects on inserts, and how their order effects processing, in more detail in Chapter 8 – Mixing in Logic.

The point to remember with plug-ins instantiated in this way is that they have no direct impact on the actual signal that is recorded to disk. While this might first seem a little frustrating, it means that, if you change your mind later on, you can always return back to its initialized state – peel back the layers of distortion, on a guitar, for example, and you'll still have the completely clean, DI'd version. In effect, these plug-ins are purely for monitoring purposes – they're keeping the musician happy with what they hear in monitor mix, and the engineer content with what is being printed to disk.

Recording through an effect plug-in will have no direct impact on what is being recorded, only what you and the musician hears in their mix.

Accounting for processing latency

Although most of Logic's own plug-in are optimized for low-latency operation, it is possible that creating long chains of plug-ins, as well as using certain third-party plug-ins, might start to add unacceptable amounts of latency for real-time recording and monitoring. If your playing starts to feel unresponsive, therefore, you might want to consider removing certain plug-ins, although these can always be reinstated once it comes to the mix (when latency isn't an issue).

As alternative to manually switching in and out plug-ins, try using Logic's own low-latency mode, which is available directly from the transport bar, or under the General Tab of the Audio Preferences (Logic Pro > Preferences > Audio), where you can also modify the accepted latency limit from the default of 5 ms. Using the low-latency mode is an excellent way of being able to quickly move between tracking and mixing tasks, without having to worry about which plug-ins are causing latency problems.

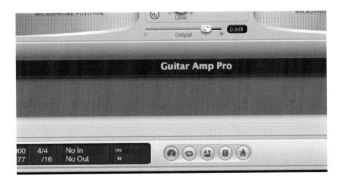

Using the low-latency mode will ensure that any DSP-hungry effects plug-ins won't create unacceptable amounts of monitoring delay.

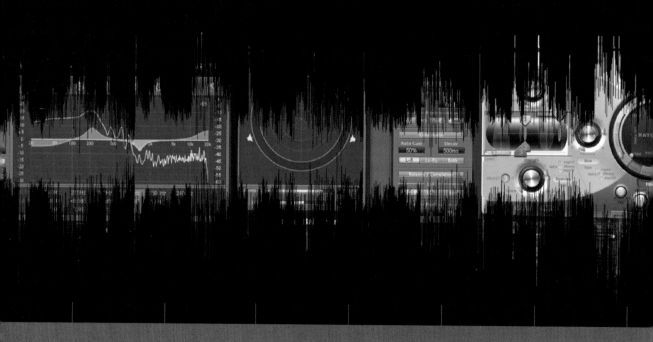

Audio regions and editing

5.1 Introduction

Editing with multitrack tape recorders was a difficult and precise affair usually employing a razor blade, and some sticky tape. Of course edits could be crude and always across all 24 tracks. Whilst this was the norm for many years in the audio profession, physically cutting and splicing two different parts together to make a new arrangement was not without its drawbacks as edits could sometimes be irreparable.

Modern digital audio workstations, such as Logic, now offer great flexibility and increased options for both rearranging and editing the parts that make up our productions. With your recording now done, you're likely to start concentrating on the arrangement and ensuring all the parts fit. Managing regions on the Arrange area is an important part of the production process, ensuring that the regions flow in the arrangement and that any edits are taken care of.

In this chapter, we'll investigate the features offered and difference between editing on the Arrange area itself and the Sample Editor. We'll learn to navigate the various windows and regions effortlessly to achieve the edits you need. First we'll start with some basic editing features and then move on to topics such as Logic's impressive Quick Swipe Comping feature, which makes Logic one of the most powerful and advanced audio editors available.

5.2 Editing on the Arrange area

Recordings are presented as blocks known as Regions on the Arrange Area. For the most part, editing directly on the Arrange area should suffice for firming up the production's arrangement and flow. However, there will come a time when the non-destructive, yet precise, editing offered in the sample editor will need to be called upon.

Editing on the Arrange area could start through the use of the traditional system-wide controls such as cut, copy and paste or employing the use of Logic's mouse Click Tools or the associated contextual menus. The editing features expand yet further due to tabs that can be introduced on the right-hand side and the editor that can be called upon from the bottom of the Arrange area.

Click Tools

For most editing purposes, managing regions on the Arrange area itself might even suffice. The myriad of features and tools available are easily called upon by accessing the mouse click Tools. To do this, simply press Escape at which point you'll be presented with a drop down menu of tools to choose from. To speed up your workflow, Logic allows you to type in the accompanying number of the tool you wish to use to get straight to the job of editing. If you prefer to click on the corresponding button to access this menu, this can be found on the top right-hand side of the Arrange page under the Lists and Media Icons in the toolbar. The accompanying tool button to the right is for you to set the Apple + Click Tool. This can really open up the speed of your editing by accessing two tools concurrently. For example, when chopping up parts the main mouse Click Tool could be kept as the Pointer Tool, whilst the Apple + Click Tool might be best set to the Scissors Tool.

The Click Tool menus can be found on the top right-hand side of the Arrange area and provide you with a wide range of editing tools.

The tools in the Click Tools menu are fairly self-explanatory and offer a great deal of control over your arrangements. They include tools such as the Scissors for splitting regions, Glue for repairing regions, Eraser for quick deletion, Solo for immediately playing whichever region you touch with this tool, Mute for muting regions, and Zoom and Crossfade, which we'll cover later. There are also some automation tools which we'll cover in Chapter 9.

As we've recently described, a lot more editing features can be found in the contextual menu for each region. This is a menu which can be accessed by Apple + clicking the region offering specific features for the management of that region. Many of these offer the user a great deal of flexibility and power when editing, many of which are covered in this chapter.

Scrubbing and editing

Editing regions will invariably involve cutting or separating regions at some point. To do this we wish to employ the Scissors Tool (Escape > 5), and point to the position we'd like to make and incision. An extra dimension to the Scissors Tool is that you can scrub across the timeline to hear where you are about to place an edit if you click and hold on the region and swipe the mouse back and forth. You will be able to hear where you wish to edit exactly. When you have found the point, simply let go off the mouse button and the region will be cut.

Another method would be to scrub using the playhead. Scrubbing refers to the action used with open-reel tape machines, where the tape would be scrubbed back and forth across a tape head to locate a particular position. Once the appropriate position is identified, the tape could be cut and the edit made. To emulate this within Logic you will need to engage pause, by either pressing the pause button on the transport bar or pressing Apple + Enter (that's the Enter on the numeric keypad, or the Enter beside the left cursor key on Mac laptops). With Logic now paused, you are able to click on the arrow at the top of the playhead and drag this in either direction to hear a sonic representation of the position. If you're planning on making a cut, select the region and then visit the Arrange area's menus to go to Region > Split/Demix > Split Regions by Playhead.

Scrubbing using the playhead can be another accurate way of splitting regions at the appropriate position.

The speed of the scrubbing can be managed by the Maximum Scrub Speed and is found in Logic's preferences – Logic Pro > Preferences > Audio. The Maximum Scrub Speed describes the speed of the audio that is played back to you when you are scrubbing audio. When this is set to Normal, the speed is at the normal speed of playback, although you have the option to double this. The Scrub Response details how fast your mouse movements will be turned into playhead actions. Given the amount of processing power scrubbing involves, the speed should reflect the power of your computer and be set accordingly.

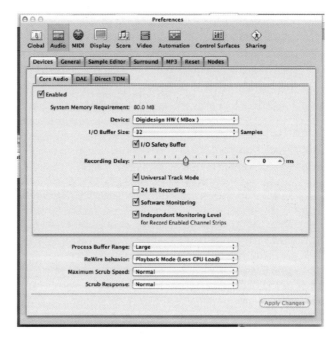

The scrub speed and the response time can be managed within the audio preferences pane.

Resizing regions

One of the most common edits made with audio is what is known as topping and tailing. This ensures that there are accurate and clean edits at the front and end of the audio passages, thus reducing on noise and other unwanted artefacts. Resizing regions is very easy within Logic and as the name suggests it is a non-destructive edit.

Regions can be manipulated easily and trimmed to fit by simply calling up the Resize Tool which appears when you point to the left or right extremes of any region. The pointer changes to an icon not dissimilar to that of a squared bracket with arrows either side. With this tool enabled, simply drag the region to the desired length, although this will be dependent on the Snap settings.

For fine adjustments irrespective of the Snap function, simply press Control after you've clicked on the region. To increase this yet further press Control + Shift which offers even more finesse.

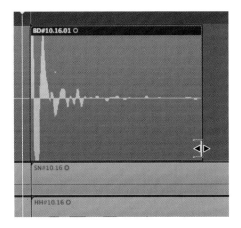

Resizing regions can be easily undertaken by pointing to the bottom left or right of the region. Note the pointer changes to a bracket with arrows on either side.

Snap and moving regions

When simply moving regions and undertaking simple editing, Logic can lock you into a grid which is related to the tempo of the project, or if working to picture, the timecode. This lock is known as Snap and can be found in the top right-hand side of the Arrange area.

At the start of new projects, the snap value defaults to its "Smart" setting. This is an intuitive way of working with snap. Dependent on the zoom level and the size of the region, Logic interprets the division you require. This is perhaps the only snap setting you'll ever need, however, there may be situations where you will wish to specify, such as simply reorganizing an arrangement where all the fine editing has already taken place. In this instance, it is quite possible that a snap value of "Bar" will be the most efficient.

Having said this, there are a number of different alternatives which may relate to the work you are undertaking and range from bars and beats to quarter frames and samples. Whether you are working with film, where timecode will

The snap menu offers a wide selection including bars and beats for composition and frames for visual work.

be your reference point, or bars and beats for composition, Logic offers all the options. There is even the opportunity to keep relative positions within bars, beats, etc. Therefore, precise, non-standard edits can remain relative to the "snapped" grid when moved or copied to other parts of the project. For example, this can be useful when an offbeat rhythm is edited to start on the up beat. Moving this part in relative mode will move it to the corresponding part of the next bar.

There will be times when you may wish to bypass the snap constraints. In this instance, it is not necessary to disable the Snap menu, but it is accessed by selecting the region and then pressing Control for finer work, or for ultra-fine movement Control + Shift. These allow for very precise edits directly on the Arrange area itself.

Nudge

At times it is difficult to move regions by the smallest of snap values and a kick in either direction on the timeline is all that is needed. Nudging offers the ability to move regions by a pre-determined amount in either direction irrespective of the snap value. This can be excellent when writing to a grid of bars and beats for composition or timecode for film work (covered in Chapter 10). To nudge parts, simply select Alt + Cursor left or right.

The amount of nudge will be determined by the Set Nudge Value feature, which is accessed by the contextual menu for that region. To reveal this

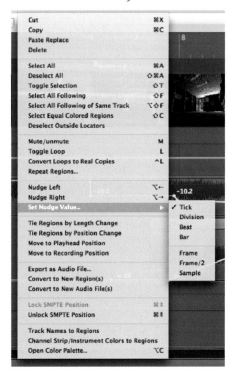

To set nudge values, open up the contextual menu for any region.

menu, simply Control + Click the region on the Arrange page. The fourth grouping of menu items relates to nudging and includes Nudge Left, Nudge Right, and the Set Nudge Value, which can be set to a number of useful settings dependent on the work you are producing. As such working with bars and beats will require that you choose the top selection of values. The bottom selections are based on working with film units such as frames and samples.

The toolbar can also sport some simple nudge controls right at the top of the screen. To select these, simply Control + Click the toolbar at the top and choose Customize Toolbar. From this huge range of possible controls look for the three buttons that correspond and drag them to Logic's toolbar: Nudge Value, Nudge Left and Nudge Right.

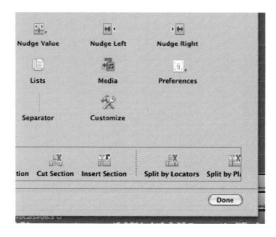

The toolbar can be edited to suit your working style. For picture work or lots of fine editing, these nudge icons might be more useful on the toolbar.

Drag

Other simple edits can take place on the Arrange page and one of the most effective solutions is the Drag selection menu. There will be times when you will want to butt two regions together, perhaps as the latter part plays later on in the take, and as such needs bringing forward.

Logic offers a number of solutions to this. Firstly, the regions can overlap, meaning that the latter part will always play and automatically mute the part below, but remains intact. This is great when you know you might want to resize the other region and leave the original audio intact below. No Overlap simply means that the first region will be edited in size to fit.

Logic also employs an automatic crossfade tool within the Drag menu entitled X-fade. Moving it forward in time without a crossfade may cause a nasty glitch as the samples change from one region to the new one. With an automatic crossfade added, the outcome would be a smooth changeover which can be managed using the Inspector which is discussed later.

Using the Crossfade Tool you can quickly swipe to smooth over transitions between takes.

The remaining two choices of shuffle L and shuffle R are important to remember as they butt together with the part you place audio onto, thus leaving no overlap and moving existing regions. It is noted here that this feature only works with Audio tracks, not MIDI or Apple Loops.

Marquee tool

On occasions it might be necessary to select a portion of a region, or indeed portions of regions for editing. For example, you may wish to make an edit in a take so that it can be repeated later in the song. Usually such an edit would require two incisions using the Scissors Tool before returning to the process of copying it to another part of the song.

Without using the Marquee tool, the whole region or regions are selected as a "rubber band" touches them. However, selecting the Marquee tool allows you to pick parts and elements within regions with one swipe of the mouse.

The Marquee tool allows for edits to take place within a region itself and can be seen by a lighter shade of grey placed over the area.

To enable the Marquee tool, press the Escape key and select it from the bottom of the list or pressing Escape > E. Alternatively, the Apple + Click Tool is set to the Marquee tool, but can of course be changed to another task for quick editing.

Marquee applies to multiple regions. After the initial selection, it is possible to choose which regions are to be involved in the edit. This is achieved by pressing Shift + clicking to deselect or reselect a region for editing. The Marquee tool is also dependent on the Snap value you have allocated and can be manipulated in the same way to create more precise selections. By utilizing Control or Control + Shift after you have begun with the Marquee tool, finer selections can be made as with Snap.

Making edits using the Marquee tool can save a bit of precious time. For example, making a new region out of the middle of another region, for muting or for copying. Choose the Marquee tool and select the area you wish to mute. Simply click anywhere within the Marquee box. The edits are made automatically and the new region is created. Next select the tracks and press "M" to mute them. This is very handy over grouped tracks also.

If your selection is not quite right, perhaps you might wish to employ Apple's new Snap-to-Transient process? When selecting an item using the Marquee tool, use the cursor left and right to extend the end of the selection to the next transient. Press Shift and the left or right cursor to access the start of the selection. Either press Alt + cursor left or right to make this a separate region, or simply click on it.

The Marquee tool can be really useful when wishing to replace a section whilst recording. Simply select the overdub you wish to make using the Marquee tool, press record and you'll be dropped in within the confines of the selection. This, combined with the Snap-to-Transient feature, makes for very quick and accurate overdubbing.

5.3 Regions: repeating, looping, aliasing and cloning

When writing music it is often good to get a feel for the arrangement through making copies of recorded regions, or indeed copying some of the Apple Loops. These "copies" can come in different varieties, which can significantly influence how the regions react when being edited. The first type would be the standard copying, which shows a straight copy with no real connection to the original.

There are, however, instances when you quickly want to build up a track and will simply choose the Toggle Loop function found in the Inspector or by simply pressing "L." Another way of creating looped regions is to select a region and point towards its top right, where the pointer arrow changes to a vertical line with a clockwise arrow. This is the Loop tool and by clicking and dragging out from this point we can create loops which can be stopped at any point.

Pressing 'L' can quickly allow you to loop a part and build up your Logic project.

These looped parts will extend through to the end of the timeline or until they reach the next proper region. Looped regions are greyed out and are, in essence, references to the original part. Therefore, if you edit the original region, all the loops will carry the same edit and follow suit. The Loop tool can be useful to reduce the length of looped regions created using the Toggle Loop command to fit your project.

As with looped regions, it is also possible to create a "referenced" copy which can be placed on alternative tracks or later in the project. These are known as Aliases and Clones and are conceptually the same thing except that Clones refer to Audio regions whilst Aliases refer to MIDI regions. Alias regions can easily be identified by their title being in italics. Cloned regions are indistinguishable from the real region.

Cloning audio regions is useful, for example, when you want to copy a chorus vocal to a few places later in the track. The benefits of cloning are that any resizing you make on the audio regions will be cloned to the others, saving you from editing each one, or re-copying. Creating a "clone" can be achieved by pressing Alt and Shift whilst dragging. This will allow for a clone to appear at a different part of the arrangement or on another track for different processing perhaps with the same file reference. It is worth noting here that resizing and cutting of a clone will be copied to all other clones of the same file.

At times you may wish to edit a certain part of your project using a clone, but are mindful that any change you make will affect the other clones. It is therefore sensible to convert this clone into a region in its own right and this can be achieved by going to Audio > Convert Regions to New Regions or pressing the key command Apple + Alt + R.

Aliases are the same as clones but are for MIDI regions and Folders exclusively and can be managed in the same way as clones. However, unlike with clones, an alias can be resized or cut without these edits being replicated on other aliases of the same region. The Alias menu (MIDI > Alias) contains a number of features such as "Make" which makes an Alias at the current playhead position. Another useful feature here is the opportunity to "Convert to a Region Copy" thus rendering this a region in its own right.

To prevent inadvertant changes to all your clones, convert to a new region.

If you prefer the more standard menu-driven approach of working, then call up the Repeat Regions… dialogue by going to Region > Repeat Regions .... From this dialogue box you can choose the number of copies you wish to create, their timeline adjustment, and whether they are real copies or cloned/ aliased regions. The timeline adjustment feature is really useful as it will quantize the start point of the next copy to the next bar or the next frame dependent on your needs.

The Repeat Regions dialogue allows you to easily specify the type of copy you wish to make and how copies should fall on the timeline.

5.4 The Inspector

As you are working with regions, there are parameters that will need editing from time to time. Whilst it is often essential and preferable to use graphical means of editing, such as the automation lanes, there will be times when a simple numerical input is required or indeed desired. To the left of the Arrange page lurks what is known as the Inspector, an important space for gathering information and making static edits to tracks and regions. This can be toggled in and out of view by pressing "I", or clicking the Inspector icon on the top left of Logic's Arrange page.

The Inspector is interactive and responds to the selected region or track by revealing important information and parameters in addition to the track's fader. The information contained here can often save time searching a mixer window or some other editor. Whilst being rather table-like, it is a veritable source of useful information about the selected region or track.

The top half of the Inspector contains a table which is known as the "Region Parameter Box" which displays information about the region, and this responds and alters depending on what form of region it is, whether it be a MIDI, Audio, or Instrument region. This area is excellent for simple data entry,

such as fade-in and fade-out times, including the curve and any sample delays. This allows for quick fine tuning where required.

This fine tuning is excellent in its ability to batch process. A large number of regions can be selected, and their fades, for example edited in one swoop, saving precious time.

5.5 The locators and cutting/inserting time

From time to time you will want to experiment with your project by making some space to try some new ideas through inserting some time into the arrangement or moving some parts. For example, you might choose that a new section should be added to the song, perhaps making way for a middle 8, or to repeat a chorus. Whatever segments you wish to make or move, Logic has some quick and powerful features up its sleeve.

To take advantage of the features in this menu you first need to draw the locators at the top of the Arrange area at the point and length of the section you wish to insert or cut. At this stage, Logic does not assume that you wish to apply this to all of the regions, but perhaps only some of them. So it is important that you select only the regions you wish to move or cut. It is worth noting that some of the regions you will select will run outside of your chosen locator length. This is fine, as Logic makes the necessary edits for you.

In the example shown opposite, it might be necessary to insert eight bars at bar 17 for a new section. In this instance we would choose the Insert Silence Between Locators feature from Region > Cut/Insert Time menu. However, if instead you wished to remove the eight bars to move them elsewhere in the project, then simply choose Snip: Cut Section Between Locators from the menu. Because you have cut this, these regions are held within Logic's clipboard and can be pasted in another part of the arrangement. To insert this part simply place the playhead position at the appropriate place and then select Splice: Insert Snipped Section at Playhead from the menu.

Once you select one of these options you are given options as to whether you wish to move global track changes or leave them

The Inspector with an audio track in focus showing both mixer strips, the left one refers to the selected track and the right one shows the associated signal path such as the group output or auxiliary track the bus send is connected to.

unaltered. Moving these events forward will allow synchronization with the parts afterwards, but there might be times when it is necessary to keep the global track changes where they are. One such example might be when some tempo changes are synchronized to a movie track.

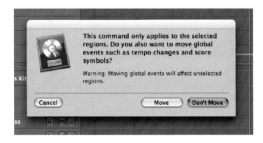

When making changes to the timeline such as inserting time, the global tracks can be either kept where they are or moved to match the extended project.

Within the same menu lurks a choice to Repeat Section Between Locators which allows you to insert a copy of a snip directly after the end of the cut section. This can be really useful when it comes to extending a chorus or middle 8 in a project. The global tracks can obviously be set to extend if you so require.

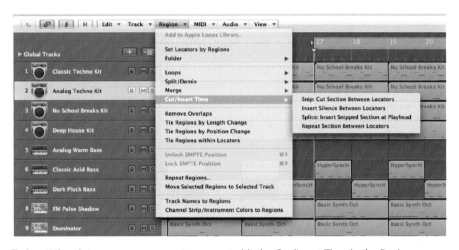

To insert time into your arrangement you must visit the Cut/Insert Time in the Region menu.

We mentioned earlier in this chapter about contextual menus and their focused set of features and tools to the task at hand. Cutting and inserting time into an arrangement is no different and the Locators hold the key to this menu. You'll notice the Snip and Splice tools included in this contextual menu below for easy access. By Control + clicking on the locators, you save yourself from an extra click stroke than going to Region > Cut/Insert Time...>.

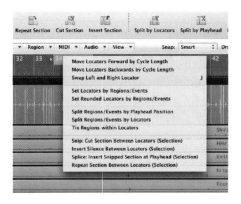

The Locators' contextual menu provides some quick access to the Cut/Insert time tools.

This contextual menu adds some very useful features such as Splitting Regions/Events by the Playhead position, or Splitting Regions/Events by Locators. Both of these can have immediate and profound benefits when editing. Simply stopping Logic at a certain point can automatically offer you an editing tool.

One major feature of Logic lurks in this contextual menu and is excellent when working on your arrangement. There may come a time when you would like to see what it might sound like to remove a section in your project. In many DAWs you would be expected to make the appropriate edit and butt the parts back up together. Whilst this is most likely a non-destructive edit, it is still relatively time consuming. Logic allows for the locators, which normally can allow for cycling through parts by pressing "C", to be swapped and therefore reject playing the material contained within. In other words, the area covered by the locators will be skipped entirely as the Left and Right locators are swapped. This can be accessed in two ways: firstly, you can swipe the locators from right to left in the locator lane; or simply press "J" to swap the locators if they're in the right position.

5.6 Strip Silence

Prior to the use of digital audio workstations, digital signal processing and of course Logic, managing spill was usually handled by a processor called a noise gate which we'll cover more in Chapter 8. The spill on another track of the multitrack caused by an adjacent microphone in a drum recording, for example, would be gated to create a more focused sound. Therefore, many elements of the drum kit would have been passed through a gate during the mix to obtain a clean, untainted sound.

However, it is possible now to edit the "silent" passages in between either manually or by the use of the Strip Silence feature. This process allows for the areas of the audio region with spill to be made silent by simply a single process. Strip Silence can be accessed in the Arrange page by selecting the audio region you wish to process, and then going to Audio > Strip Silence, or pressing Control + X.

As with a traditional noise gate, a number of features exist to tailor the response to the audio file. Strip Silence offers a threshold, an attack and release time as well as a feature called Minimum Time to Accept As Silence. This is a useful feature to prevent machine-gun-like results as the Strip Silence feature works on a fast drum fill or similar. This, in essence, smoothes over the results so that if two "hits" are close enough, then the process will ignore them and ensure a smooth response.

Using Strip Silence is quite intuitive and visual results can immediately be seen on screen. As you increase the threshold, less boxes will appear around the blacked out waveforms, meaning that less of the original audio will be heard. Logic keeps a greyed out version of the waveform available to see if you are missing any important parts of your take.

Zooming within this view is not possible and consequently it might be very difficult to see the proposed edits on a large region. As such, you may wish to limit the size of the region you wish to process to a manageable section, such as the chorus, verse, or middle 8, thus ensuring you can see the detail required within the Strip Silence viewer.

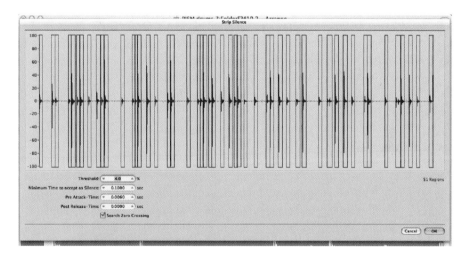

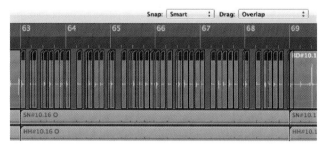

Before mixing it is often desirable to remove the noise and spill in the background. Strip Silence can make easy work of this within Logic. The result is a non-destructive set if regions placed back on the arrangement.

Once you're happy with the settings and the edits appear correct, simply press Ok and wait for Logic to process it. The result will be a number of small regions on the Arrange page which correspond to the "hits" the Strip Silence feature found. This is a non-destructive process, so you can simply resize any of the regions if a hit is either missing or cut off abruptly. A benefit of these parts now being discrete "hits" is that you can choose the best one and place it in the timeline as you see fit. This opens up the possibility of tightening up the performance both in terms of the sound itself and the timing by moving parts to the appropriate place. Once parts have been stripped it is possible to quantise the remaining parts. This is achieved by going to the events list, by selecting the Lists icon to the right of the menu bar followed by the Event tab. Within this page resides a list of the various quantisation values. Choosing one of these will shift the selected parts into time.

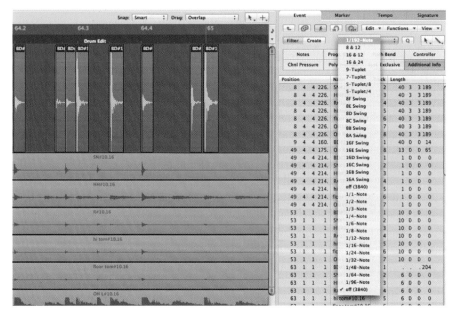

Strip Silenced parts can then be quantised using the Event List.

5.7 Quick Swipe Comping

Taking multiple passes of an instrument to obtain that perfect take is a common process of studio recording. At some point there will come a time when the decision needs to be made which take or elements of a take should be brought together to make the new performance. Not so long ago, this common process of compiling separate takes together to make a Composite take (known in the business as Comping) was a laborious one and at times both tricky and skilled. Whether you used an analogue tape recorder or some early digital audio workstations it could be nail-biting at times.

Comping in Logic now makes this process ever so easy for single tracks at a time like vocal takes. Editing this way can be used for so many purposes and with Logic's new Comping features, it is possible to experiment and carve out the perfect take for your project, leaving every variation intact should you want to come back to it. With this speed and flexibility, your perfect take can be realized in a few moments.

In Chapter 4 we introduced the notion of cycle recording and Logic's ability to keep these takes for you for future editing together. As you record on cycle, or if you just record again over a previous take, Logic keeps these for you in the Take Folder menu which can be accessed by the small arrow facing downwards at the right-hand side of the region. If you wish to select another take into the arrangement replacing the most recently recorded one, simply choose an alternative starting with "Take:" from the menu.

Nevertheless at times you'll want to take the best parts from each take and begin the process of Comping. Logic's Quick Swipe Comping feature allows you to quickly choose the parts you wish to play from all the variable takes you may have recorded. To get started, click on the small arrow facing to the right at the left-hand side of the region, or double-clicking the region as you would with a folder track. The arrangement expands to show you all the takes you have made for this region upon the one audio track.

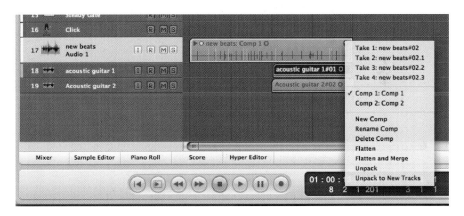

Choosing an alternative take you may have recorded is simple using the Take Folder menu.

Logic's ability to compile these together is impressive. Once the takes are expanded Apple's Quick Swipe Comping can be employed. You simply swipe the sections of any take you want, and they'll play only those swipes! It is worth noting that Logic imagines that you wish to use the last take as the default unless you swipe an earlier one, so there is always audio playing. Simply swipe an alternative you think might be better than the one you're listening to (the last take) until it all gels together.

Logic's Quick Swipe Comping feature enables extremely fast
and accurate compiling directly from the Arrange page.

There may be a possibility that you wish to try some silence in places. For
example, it might be desirable to take out the extra "ooh" or "yeah" from the
vocal take. To do this simply press Shift as you swipe and the adjoining region
will not be extended to fit, thus leaving you with silence as the last "default"
take will not play.

For most applications, each crossfade will perhaps not require individual
tweaking. As such, the Quick Swipe Comping Feature has a fixed crossfade
time and curve. These can be altered in a generic fashion before you apply
any swipes. To alter these, navigate to the "General" dialogue box within
Logic's main preference Logic Pro > Preferences > Audio. The last two sliders
here allow you to alter the crossfade time and curve.

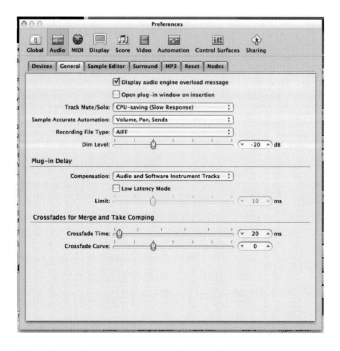

The Audio Preferences pane is the key to the finer management
of the Quick Swipe Comping feature crossfades.

If you are happy with your comp, but would like to experiment with alternatives, then select the Take Folder menu triangle to the right-hand side of the

region and select "New Comp," you will notice that under the take list there is a list of Comps too. These are your different attempts and offer you non-destructive ways of putting the take together. It is, as always, advisable to take the time to label the comp using the "Rename Comp" tool from the drop down menu, if only to signify the comp you like the most.

At some point, you may wish to commit your selection to the arrangement and there are a few ways in which this can happen. The first option is to "Flatten" the takes. This can be found in the drop down menu. In this selection, the edits you make using the Quick Swipe Comping feature become audio regions in their own right (as shown below). This can be useful for all manner of things such as rearrangement, muting, copying and looping of parts.

The other method available within the menu is the "Flatten and Merge" option. In this scenario, the edits are bounced to make one new audio file, as though you were printing back to tape. This allows for further editing and treatment as an audio file in its own right.

When you flatten a Quick Swipe Compilation, each segment is butted together (unless you have chosen otherwise using Shift) and an automatic crossfade is added.

Unfortunately there is a small downside with Quick Swipe Comping which requires a slightly different approach. The takes within the expanded folder cannot be moved against each other to remedy any timing issues. As such you would need to take a more traditional approach such as expanding the takes to multiples of the same track to edit. To do this simply choose the "Unpack" option from the Take Folder menu and this expands the takes to multiples of the same track.

Alternatively for small timing alterations you could start with Quick Swipe Comping to get very close to the best take. Next, flatten the track for further editing as separate regions. It might be possible to move parts with crossfade selected from the Drag menu to re-align the passage.

The Unpack to New Tracks option allows for you to spread these takes to new discrete channels for more flexibility. Having these takes on different channels means that more than one take can play at the same time, which might over-come the timing situation above with Quick Swipe Comping. This might also be useful for all sorts of things, such as wishing to use two takes as a double-tracked lead vocal, or some backing vocals when retuned.

5.8 Tempo-based editing

Adding new parts to projects can turn up some unexpected frustrations when the parts don't quite fit because of a tempo difference. In the old days it would have been a laborious off-line process to match up the speed of the imported audio to the tempo of the project. Similarly when working with live musicians, their performance does not always match the clinical tempo of the metronome click making editing slightly more difficult to manage. Nowadays, Logic has some impressive methods to manage situations such as those which we'll explore in this section.

It might be necessary to import some audio, such as a great loop into your project which is not at the same tempo. Ideally we would wish to time stretch or compress these parts to fit the tempo. Logic manages this in some clever ways. For example, if you drop in a piece of audio which is too slow (long) for the bar you have in mind, you have the ability to compress this to fit. If it is really close to the length it needs to be, then choose Audio > Adjust Region to Nearest Bar which will compress it or expand it to fit.

A simple and quick way of working is to simply select Option and drag the Resize tool at the end of the part to the desired length. The region will then be changed in length and will be governed by the current snap settings. There is another way by making the locators fit the ideal length you wish the audio to become. By then selecting the region, it is possible to apply the time compression to this region to fit the locators. To do this go to Audio > Adjust Region Length to Locators. Once processed, the part will now fit the desired length at the same pitch.

We've established how to make the audio fit the project tempo using time stretching or compression, but there may be a piece of audio whose tempo you wish to base your project, or new section, on. In this instance, Logic is able to determine the tempo of the audio you have provided by matching length of the region to the locators.

Simply set the locators to the length that the region ought to be, say two bars. Next, select the region and either press Apple + T or go Options > Tempo > Adjust Tempo Using Region Length and Locators. You are then given two options, whether this treatment is to last the whole project or just the length of the region. This can be useful for all kinds of things, from film composition, sound design through to middle 8 experiments within your project.

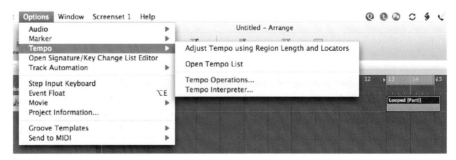

Basing a Logic tempo upon a region is simple by using the locators and the Adjust Tempo using Region Length feature.

Follow Tempo

In the past, should the producer or band feel that the song, or passage, they have just recorded is a little slow, then the only options available to them would have been to speed the multitrack tape machine up a notch, so the output would be faster and higher in pitch, or to simply re-record the whole track again.

With natively recorded Logic Audio Files, or files that Logic has created either through exporting or bouncing to the Audio Bin, tempo changes in the project are automatically acted upon. If you want to "speed it up," so be it as Logic will ensure all the appropriate time stretch/compression is taken care of. To allow Logic to take care of this, you need to click the Follow Tempo check box in the Inspector.

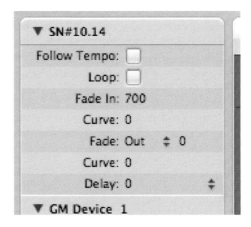

Click the Follow Tempo checkbox to allow Logic to play this audio file at the corresponding tempo whenever it changes.

It is worth noting that this feature only happens in the original project. So new recordings will be manipulated when the tempo changes. This is because Logic can record the tempo it was played at originally. Imported tracks will need to be either exported or bounced to the Audio bin to take advantage of this feature, so that a reference tempo is associated with the file within the current project.

Beat Mapping

To click or not to click, that is the question? Using the metronome to provide a click track to the band is a common aspect of recording these days. It keeps the band to the tempo of the song and makes editing the audio at a later date much easier. However, for some bands the click will simply not do as it can change the way in which the music flows for them, or it is simply off-putting.

However, Logic offers a powerful solution to this problem! Within the Global Tracks Configuration dialogue lies a check box called Beat Mapping. With this checked, a new lane appears in the Global Tracks called Beat Mapping. When this is expanded, a new set of divisions is shown at the top of the lane which correspond to the bars and beats as set by the project tempo and time signature. The bottom half of the lane is reserved for the audio or MIDI region from which Logic adopts its tempo information.

Beat mapping allows you to follow the tempo of an audio or MIDI track. It manages this by showing the selected region's rhythmical information in the Beat Mapping Global Track lane. For MIDI tracks, the actual pitches are neutralized and only the rhythmical information is shown. Audio region waveforms are shown, but need to be analysed before they can be mapped to the tempo.

When matching audio regions, first we need to select the region we wish to use as the reference. Typically this will be a drum part, perhaps the bass drum or a high hat track. These tracks are usually the easiest parts of the kit to get the required information from, because more often than not they play the straighter aspects of the rhythm. Nevertheless, other off-beat parts could be used and mapped accordingly.

On the Beat Mapping track header lies an Analyze button which needs to be used to identify the transients in the waveform, which will then translate into a tall line in the lane. There will be instances when you will try to analyse potentially louder passages and too many of the transients may be picked

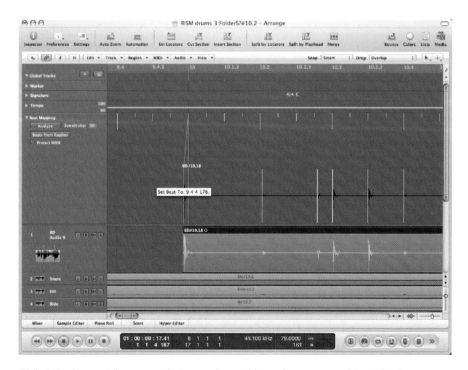

Global Tracks provide a neat solution to the problem of tempo matching. The Beat Mapping function in Logic allows you to link up the tempo to a wavering drummer.

up. In this instance alter the Sensitivity to reduce the number of lines on the lane.

Next, the analysed lines need to be matched up to the tempo in the timeline. Simply click on a beat on the timeline at the top and connect to the corresponding transient marker. Working through the whole of a take can be a laborious process, but will enable Apple Loops and other recorded Logic part to expand or compress automatically to fit the tempo at any part of the song.

At times it may be necessary to map a beat to a missed beat or transient. To overcome this, click on the appropriate line in the timeline. A new yellow line should appear. The position of this can be moved by Control + Shift + Click and dragging the line in either direction. This can be quite a shot in the dark, but with care can produce the desired tempo change you need.

There may be occasions when you will receive tracks to mix which are perhaps from another DAW package but include a MIDI-click track. This MIDI click can be used as a tempo map through the use of the Beats From Region button. Simply select the region and click the button and Logic will map the tempo to match the beats of the region.

Snipping

Many great tracks over the years have manipulated instruments in such a way that the audio was muted on and off in a rhythmical fashion. This was often achieved using key gating, where a gate was placed across the desired sound, and then a rhythmical track, such as a high hat would be routed to the inaudible key input. Within key gating, the external key controls the gate's response and as such the output of the gated sound. This can create a pulsating sound which can be desirable to the drive of a track.

Logic takes an interesting and fast approach to this. Using the Scissors Tool (Escape + 5), will allow you to cut the region at the appropriate place, although by appending Alt whilst using the Scissors at the start of the track will produce the opportunity to "snip" the region to 16th, 8th, or whatever division you need. It is then simple to go through each new small region using the cursor keys and press "M" on alternative regions to re-create key gating with both MIDI and Audio Regions.

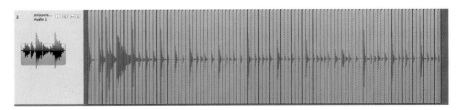

Using the Alt key whilst cutting a small section at the start of a part will continue the process to the end of the region.

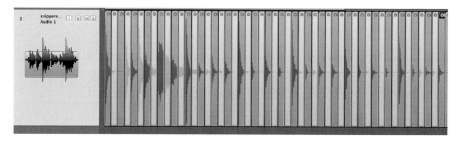

These can then be muted every other one to create an interesting key-gated effect.

5.9 Editing multiple tracks

In modern music production it is very likely that you will wish to edit multiple tracks at the same time. Cleaning up these takes and the audio is a central process of the work you are likely to do in Logic. Taking raw takes and editing them into shape can take a little time, but Logic has some clever tricks up its sleeve to help you along with this, and by employing the menus we can get started really quickly.

There are many ways to get into editing multiple tracks; one obvious way is to utilize the Marquee tool which, as we have already mentioned, allows you to choose parts of regions across many tracks for editing. By using the Marquee we can make cuts, mutes, or pretty much anything found in the Click Tools menu on a multitude of takes spanning many tracks. For the odd edit here and there this is an excellent feature. However, having to select the parts in the manner for each edit would become time consuming.

Click Tools allows for the Apple + cursor to set up an alternative tool such as the scissors for rapid editing.

Another way would be to pack the desired takes into a folder group which can then be edited and then re-organized across the arrangement. Although perhaps the most suitable method would be to group the tracks together and edit on the Arrange area.

Edit grouping

For more longer editing sessions it might be beneficial to group them as described in Walkthrough 1 on p. 110. This is achieved by opening up the Mixer Area pressing "X" or the Mixer Window (Window > Mixer Window or Apple + 2) and clicking on the Group Slot area which can be found directly above the automation mode, which is usually set to "Off" as default. Clicking within this area will show a long list of groups and features. As described in the walkthrough, we can select the aspects of the tracks that shall be grouped using the Groups Settings.

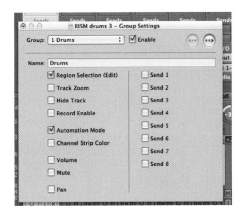

Setting up groups can be a very flexible way of managing your projects, especially when editing across multiple tracks.

Logic's Group Settings box offers many different ways of managing tracks together. In the context of Edit Grouping as we're doing here, it is important that the Region Selection (Edit) is selected as this will allow us to simply choose or edit one region and for it to be mirrored across all the grouped tracks. If used in conjunction with the Marquee tool, the portion of the region you have selected will also be matched across all the grouped tracks. This can be really powerful when editing parts of tracks together. The best of both worlds!

There are other useful features to be found within the Group Settings dialogue box, which might be of use to you when you're editing. For example, the Track Zoom link allows for all the track lanes to be expanded by the same margin automatically when editing. To expand track lane height simply point towards the bottom left-hand side of a track's header to the left of the track number. You should notice that the pointer changes from a hand to a finger. Click here and drag down to expand the track lane height or upwards to reduce it.

Other useful features lurk here when grouping, such as the Hide Track feature can be linked across the group. We'll cover this later in this chapter. This enables you to quickly hide the grouped tracks to concentrate on something else. For example, on a large production including a full band plus some orchestral work, it would be beneficial to group the Hide Track feature for band so that you might concentrate on the orchestral elements.

The Channel Strip Colour link is useful as it allows the colours of the group within the Mixer to be duplicated for easy identification. Again many of the other selections within the Group Settings dialogue could be of benefit in the grouping when editing, such as mute and many of the remaining features are ideal for other activities such as mixing and are covered in an appropriate chapter in this book.

The colour of the base of the channel strip can be grouped, making identification easier.

At times it will be necessary to simply alter one track individually away from the group and in this instance Logic employs something called a Group Clutch which will disable all groups To edit independently of the groupings, the Group Clutch can be enabled by going to Options > Group Clutch or by pressing Apple + G. This is a toggle switch so another press of Apple + G will result in the groups being re-enabled.

Working with multiple regions

With the tracks now grouped, editing multiple regions is easily carried out. Whatever you wish to do, select one track and the whole group will be selected. Choose a tool and click on one of the regions, and the whole group's regions follow suit. This makes editing work very fast on multiple tracks for resizing or indeed anything that can be found in the Click Tools menu.

For example, it might be beneficial to rename all the regions, not by their track relationship but by what section they represent. In this instance, we'd choose the Text Tool (Escape > 4) and click on one of the regions. By changing this to "Chorus" or "Verse" will offer you a quick indication of where you are in the arrangement. In the example below, we've called this "Middle Eight" and all the tracks have been renamed across that grouped section leaving the original names for the regions that follow.

Cutting regions will be an important part of the grouped editing undertaken in projects. Using the Scissors Tool will be a common feature of this process (Escape > 5) and will enable you to cut across all the tracks at once making rearrangement of your project very simple.

Renaming grouped regions using the text tool can be quick and useful for easy identification of their contents.

However, this is slightly different when you select the grouped regions and choose the Glue Tool (Escape + 6). Logic presumes that you wish to glue these tracks together as opposed to the adjacent part and as such translates this to mean a mixdown of the parts. As such you are presented with a dialogue box which enables you decide whether you want a mono or stereo mixdown.

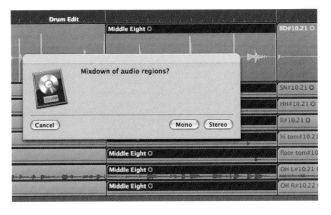

Gluing grouped regions together results in a mixdown which can be either mono or stereo.

Walkthrough 1

Grouping Multiple Drum Tracks

Step 1:

Working with large track counts is a key part of modern music production. Having the ability to edit across all these tracks in the Arrange Area is an important first step. To do this we need to group the tracks together by choosing the Mixer from the Arrange page by pressing "X". Next, navigate to the "Group Slot" on the first channel strip. Click here and a menu should appear opening up the groups and settings. Turn your attention to the Open Group Settings … within the Group Slot menu.

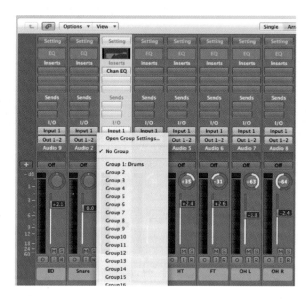

Step 2:

A new dialogue box will open up showing the settings for the first group. Simply name the group and choose the most appropriate settings. In this case we've called the group "Drums". When you create a new group, the Automation Mode, Volume and Pan boxes are checked. It is worth deselecting the Volume and Mute boxes at this stage as we're only interested in editing at the moment.

Step 3:

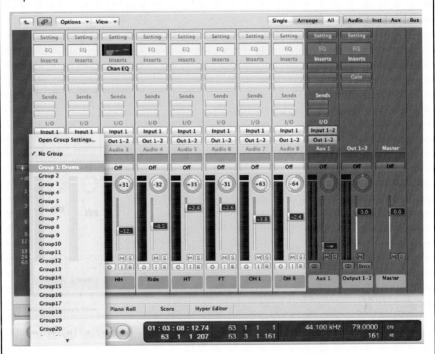

Next we need to apply the "Drums" grouping to the remaining tracks. To do this click on the Group Slot on each channel strip and select the appropriate group. If the drum tracks are adjacent in the mixer window, a quicker way is to select the leftmost channel's name, then click and drag right to highlight all the tracks to be grouped. The tracks will turn light grey in colour. Now select the Group Slot and choose the Drum Group. The drum tracks are now grouped and ready for editing together.

5.10 Region management

With so many regions on the Arrange area at any one moment, it is often difficult to see the wood for the trees. As such Logic takes its management of regions and your session seriously. Therefore, it offers you some excellent tools to assist your workflow from grouping regions together into folders to hiding tracks you no longer need to edit, thus leaving you with greater screen real estate for the task in hand.

Hide tracks

Track counts in today's productions can become much larger these days and it is often difficult to focus on the tracks and regions you are specifically working on. Hiding tracks that are not in use at the moment, or muted for consideration later, can be taken out of focus using the Hide function.

On the Arrange area to the top left lurks a button labelled "H," which is known as the Global Hide View Button, and when depressed a series of buttons also inscribed "H" appear on each track beside the record enable button. Simply choose the tracks you wish to hide and then click on the "H" button at the top of the page. The tracks you had selected to Hide will now be out of view and will improve your ability to focus in on the current important tracks and regions.

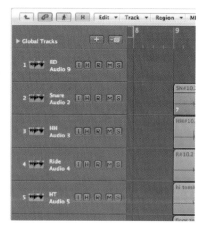

The Hide tracks button can be viewed in the track header by pressing "H" on the keyboard.

Folders

In modern music production it is common to have many tracks dedicated to drums, vocal takes, backing vocals, or even an orchestra in some cases. Folder tracks enable you to pare these down to manageable regions which save on screen space, make editing sections easier and improve workflow.

Other features of folder tracks include the easy reorganization of the order of the parts in the arrangement. Folder regions offer a quick and simple solution to manage this process by making a track which can be treated the same as you would treat a MIDI region and edit it accordingly using the Scissors and trim features. For example, you could pack up folders for each part of your arrangement: verse; chorus; middle 8, etc. With these parts packed, it is then possible to reorganize the arrangement easily without needing to fiddle about with rubber-banding over a large amount of regions.

A folder could be used for the drum tracks used in the walkthroughs. Often these can span many tracks and perhaps be unwieldy in the arrangement of the project. To create our folder track, we need to select the drum regions we wish to add to the folder. In this instance, we can either rubber band the regions we wish, or we could click on a track on the track list, noticing that it selects all the regions on this track. Next, pressing Shift and clicking on all the other tracks required for the folder, you can ensure you have selected all the parts. To then create the folder, go to Region > Folder > Pack Folder.

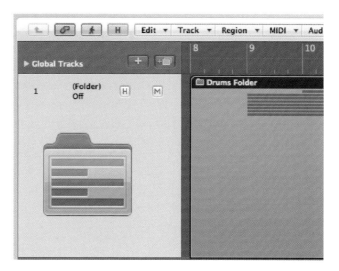

Folders can make sense of large track counts such as those for drums when editing.

With the new folder packed, it will appear on the Arrange page as an add-itional track labelled "Folder." You should rename this by double-clicking the folder name in the track list, especially if you are going to pack lots of other instruments into folders too. Now that this is done, we can go ahead and chop up the arrangement and move parts as we see fit to improve your project.

Invariably, you'll need to go into a particular track within that folder to make an additional edit. To gain access to the folder simply double-click on the folder region itself. The folder expands to show the contained tracks within. To go back, there is no need to repack the folder, simply press the Hierarchy button which goes back one step, similar to the back arrow on an Internet browser or the Mac's Finder.

 The Hierarchy button enables an exit from the folder view back to the arrangement.

As the editing progresses, there is a likelihood that you will need to either add or remove regions from the folder. To add to the folder simply drag the required regions on to the folder track. Removing tracks from a folder is a little more involved and can take two forms. The first would be to simply open the folder track and use the cut tool (Apple + X) and paste it back in the Arrange area at the playhead position. Alternatively, open up a separate Arrange window by visiting Window > Arrange Page or pressing Apple + 1 and paste the part from the opened folder to the other Arrange page.

5.11 The Sample Editor

So far in this chapter we have discussed non-destructive editing directly on the Arrange area for most of our edits. For most work, this level of editing will be more than adequate and we can achieve a great deal this way, but at some stage there will be a need to engage in some more precise and detailed editing. To do this, we'll need to look at the Sample Editor.

Regions in Logic are simply references to the audio files themselves. This is why they remain non-destructive and can be edited, resized and copied without fear of losing the original source material. At times it will be necessary to alter the audio file that these regions refer to, thus rendering any "references" also altered. Editing using the Sample Editor is accurate and flexible, but can be a destructive process meaning that only the undo feature can repair any poor decisions. This can only be achieved for the preset number of Undo steps in the preferences (Preferences > Global > Editing). The Undo History dialogue can be accessed from Edit > Undo History... or by pressing the Alt + Z key command.

Logic's Sample Editor is instantly accessible and lurks at the bottom of the Arrange page alongside the other editors. It can be revealed by either pressing "W" on the keyboard or clicking on the Sample Editor button above the transport bar. With the Link button selected, any region clicked on the Arrange Area will be portrayed within the Sample Editor.

In the example below, a reverse cymbal is shown in the Sample Editor. The audio file can be viewed clearly alongside some other useful features, which

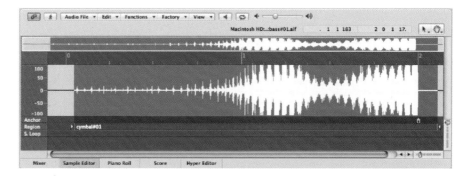

The Anchor point on this reverse cymbal has been placed at the crescendo, which will now snap to the beginning of any bar it is placed on.

we'll be exploring in this section. The first is Anchor. Anchor is the reference point for the audio region. For example, if you reverse a cymbal, the decay portion of the cymbal will now be before the attack and it will still be necessary to anchor the attack point to the tempo of the track. In this instance, the Anchor point could be moved to the original "attack" portion of the waveform. Whenever this is placed in the arrangement, it will now naturally place this Anchor point to the grid of the song: in this instance, the beginning of bar 2.

Below Anchor, is the Region length, which can be edited to suit. This will be the length of the audio played on the Arrange page. This has many excellent benefits, namely to edit the correct audio to length without losing any potentially important material. Finally, there is something called "S.Loop," which stands for Sample Loop. This allows you to edit the loop points should you wish to use this audio file as an audio loop within Logic's sampler, EXS24. We'll explore this a little later.

For most edits, using the Sample Editor is simple and intuitive. Just select the area you wish to make a change, then look up the related task in the menus above or choose the keyboard shortcut. For example, it might be necessary to make a crude deletion to the start of a file, prior to the guitar starting in the example below. In this example we have selected the initial noise prior to the guitar starting. Notice that the Anchor point is still at the very beginning of the part (bar 1). To delete this section simply go to Edit > Delete or press Backspace on your keyboard. Logic warns you that this is a permanent edit to the file, but this can be undone at this stage only by choosing Apple + Z or Edit > Undo.

The problem with this kind of edit is that it will automatically butt the audio from bar 3 now to bar 1. This is undesirable, as this part is in time with other tracks on the Arrange page. As such to employ the region area tool would be preferable. To do this, move the Anchor point to the "reference point" in the part, in this case bar 3, and resize the region area to start at the Anchor point.

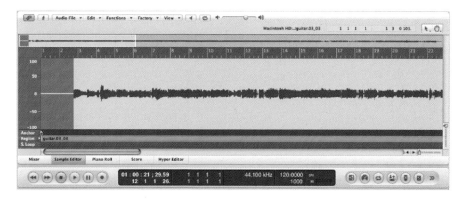

The Sample Editor can provide accurate and powerful edits of audio files.

But first ensure you have enabled the "Compensate Region Position" (Edit > Compensate Region Position) feature, which means that audible content stays in the same place on the arrangement. This perhaps seems a little overkill, when a simple region trim on the Arrange page might suffice.

Most editing can take place right within this window on the Arrange page. With the "Link" feature enabled, it is easy to switch from audio file to audio file and edit away. Navigating between windows is easy too. In Logic, the active pane in the window is usually slightly lighter shade and by using the Tab key, you can bring each pane into focus. With the focus on the Sample Editor, five menu bars can be seen (Audio File, Edit, Functions, Factory and View). These menus hold the key to an expressive amount of editing tools. Most explicit are the Functions and Factory menus.

Functions and Factory menus

Taking a look at the Functions menu first shows that many of the main features for editing audio are accessed here, such as normalize, reverse and fades. Additionally, it is worth noting that many of these can be accessed using keyboard commands (e.g. Normalize is Control + N) as shown in the menu itself. Most of these processes are permanent and whilst they can be undone immediately afterwards, it is worth noting that copies of any audio files can be beneficial.

The Sample Editor's Functions menu provides many important editing functions.

If a file is selected for editing that has a number of clones across the arrangement, any process made, such as reversing, will take its effect across the whole arrangement. As such it is wise to copy the audio file before editing. To do this go to the Audio File menu in the Sample Editor and choose Save A Copy As ..., or if it is only a part of the audio file you wish to edit, then Save Selection As ... (Apple + Alt + S).

The Audio File menu within the Sample Editor offers
some crucial management tools when editing.

With this discrete file now saved, you should consider editing the newer file,
thus leaving all the referenced copies in their rightful place. To do this, call up
the Audio Bin using "B" and drag the new file onto the arrangement at the
appropriate place, select, and open the Sample Editor to process.

The processes found within the Functions menu can be applied in other ways
that are not destructive using plug-ins and other editing features. However,
aspects such as Inverting the phase, reversing the DC Offset, or Normalizing
could be considered as a permanent change to the audio file prior to mixing.

The Factory menu provides some unique legacy aspects found in Logic, such
as the Time and Pitch Machine, Groove Machine and Audio Energizer. The
Time and Pitch Machine enables some excellent time-stretching and pitch-
alteration algorithms. These features are still very useful and may indeed be
called upon from time to time. However, due to Logic's flexibility with tempo
matching on the Arrange area, it is less visited these days.

The Factory Menu provides some useful tools,
although many of these are legacy features
which can be performed using plug-ins or other
techniques.

The Audio Energizer is essentially a simple form of compression which can be applied to the sample permanently. Some sliders allow for alterations to the attack and decay of the process. For some simple noise reduction and spike detection, the Silencer can be used to clean audio files up.

5.12 The Audio Bins

When you begin a project, the audio you have captured will have been saved within the project. This should be a bundle of audio files which can be seen in the Audio Bin tab to the right-hand side of the Arrange page (press B to toggle the Bin in and out of the Arrange page, or Apple + 9 to produce the separate, and slightly different, Audio Bin page). The audio files relating to the project are found here and offer a central place to browse, preview and do some rudimentary file editing. Basic information about the audio file is shown such as the sample rate, bit depth, mono/stereo information and the file's size.

In the Audio Bin tab, the files contain small triangles to the left of them. This indicates that each item is expandable and when clicked on opens up some information about the region's length in relation to the total audio in the file. Each window below the title shows the length of the audio used in the region by the audio file. For example, when you click the record button, Logic instantly records, but does not show you this on the arrangement. The audio region begins at the point after the bar's worth of pre-roll. The coloured elements within the long window roughly inform you how much of the audio file is used in the region.

The Audio Bin window

The Audio Bin window, as we have established, is slightly different to the Audio Bin tab found within the Arrange page. The Audio Bin window can be called up by either visiting Window > Audio Bin or pressing Apple + 9. Employing a separate window will be of benefit when needing to see more information about the waveform and some editing. For example, the Audio Bin window offers the ability to trim the audio within the region and adjust Anchor points to suit. It also shows some additional information about the audio files such as each file's location on disk.

Additionally, the file's waveform is more defined in the separate Audio Bin window and Logic offers users the ability to edit the Anchor point directly from the Audio Bin. This is achieved by dragging the dark arrows to the base of the wave viewers. Zooming within this window follows the same set protocols (Control + cursors) for most editors and as such it is easy to obtain a very accurate positioning of the Anchor point.

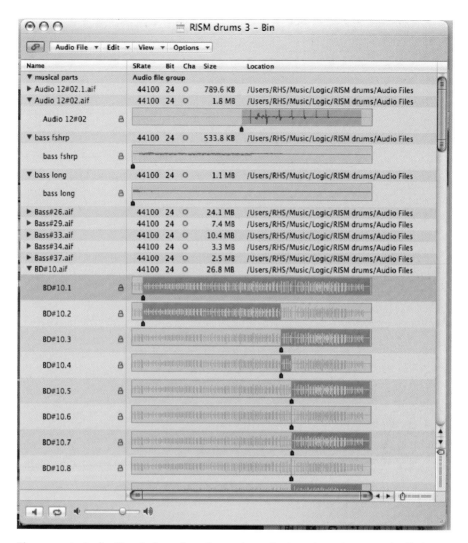

The separate Audio Bin window allows for anchor adjustments on the more detailed waveforms.

Some menus available here are Audio File, Edit, View, and Options; although the Audio Bin tab omits the Options menu. Some useful features reside here to make your file management more effective. A useful example of this is the ability to group files in the Bin. First they can be grouped by either their location, file attributes or Selection in the Arrangement. To do this, visit View > Group Files By.... Navigation of a full 24-track (or larger) session with all the various takes and extra parts can be quite long. However, using the Create Group feature (View > Create Group or Control + G), personalized groups can be created to manage and locate say, the drum files, or the vocal takes and so on.

Optimizing files in the Audio Bin

With large projects perhaps running at 24 bit, it is likely that a large amount of data is lurking on your drives. Logic offers its own optimization tool to evict any unused audio within the files that are outside of the current region boundaries and any files not used in the arrangement. To access this tool we must first select our tracks to process, which might be single parts or all of them. To select all of the files, press Apple + A, or Edit > Select All. To optimize the files, go to Audio File > Optimize File(s) ... and a new dialogue box appears which informs you of the total samples that are to be removed destructively from the selected number of audio files.

The Audio Bin tab is part of the Media lists in Logic and is very useful, although for more power and flexibility it might be necessary to call upon the separate Audio Bin window.

Logic enables you to reorganize how you see the audio in the bin, and also allows you to personalize the management of the files by the Create Groups option.

5.13 Making Apple Loops

Apple Loops are an excellent place to start working with Logic for inspiration and a starting bed for ideas. As these ideas flourish and unravel into compositions of your own, there will come a time when the audio you have recorded could become something you might wish to keep as a loop for future use. Therefore, making Apple Loops can be a beneficial way of getting you started. This will save your time on the type of beats or synths you like, for example.

Saving audio as Apple Loops has the benefit that should you use these loops in new projects, they can automatically follow the new tempo and key. Additionally, because of the additional data you have to add to any Apple Loop creation, searching can be fast and rewarding.

Saving audio as Apple Loops is simply a matter of saving the chosen audio region as an Apple Loop by going to Region > Add To Apple Loops Library ..., where a dialogue box will come up showing some of the information that can be edited with respect to this file and some descriptions about it. Broadly, Apple Loops fall into two categories, one of which is loopable information which matches what is called the "project tempo" and the other is "one-shot" samples, where the sound might be a simple hit or something such as a sound effect and is not related to tempo.

When making Apple Loops this way, it is presumed that the audio fits perfectly with the current tempo of the project and works on this basis. Each transient "hit point" is therefore set and cannot be manipulated to fit the audio in any way. Should we need to do this, we need to engage the Apple Loop Editor.

The Add to Apple Loops Library dialogue box showing some of the more basic metadata that can be added in before creating new Apple Loops.

The more flexible and advanced method of making Apple Loops is also very easy to navigate using the standalone Apple Loops Utility programme. Calling this up from within Logic can be achieved by selecting the region or regions you wish to use and visiting Audio > Open In Apple Loops Utility .... A dialogue box will often appear asking you to select its length. This allows you to specify how long the loop is irrespective of the region size or project tempo by setting the length, or if the region does not match the project tempo, but was supposed to, you can shorten the loop length to fit to the project.

Logic places an additional dialogue box before exporting to the Apple Loops utility in which the length of the audio can be described.

With the lengths of the source material determined, the Apple Loop utility automatically launches and presents its first tab which is called Tags and is essentially a metadata dialogue box. By metadata we mean simply "data about data" which will tell the Apple Loop browser what the file is and what genre, tempo and style it is. Other aspects within this box include the opportunity to add the author details, copyright and additional comments about the loop.

The Apple Loops browser contains some intelligent search facilities, and the data that are entered into the Apple Loops Editor's Tags tab will inform Logic about what loops to look for. Making sensible descriptions of the loops at this stage will in essence make searching for it again far easier. Additionally, you can specify what the loop is and how it works in any future Logic arrangement. Things such as how many beats it is, the scale, and time signature will later inform Logic what treatment the loop needs to fit your new project. The rhythmic element also informs what happens in the second tab called Transients.

The Transients tab in the Apple Loops utility allows for the tracking of the transients to which the loops can rely. Apple describes the power of Apple Loops as being able to match any tempo and key. The Transients pane is therefore important as it allows any loop to be placed within any project and be automatically adapted in real time. Choosing the transient division is important to inform the Loop utility where the main hit-points are. The sensitivity of the beat detection can be adapted to find quieter transients within the waveform. As transients appear, the markers can be either deleted or moved in the timeline to fit the loop.

With the loop completed and "tagged," it can now be saved and placed in storage for a later date. Apple Loops need adding to the loop directory by simply dragging from the Finder into the Loop browser within Logic.

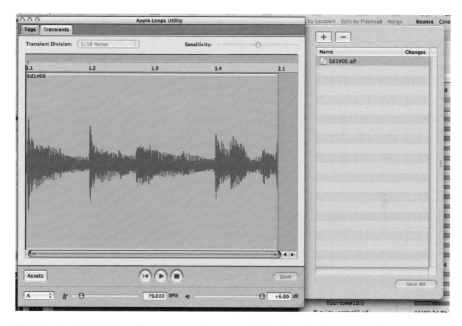

The Transients tab allows for accurate placement of the beats for future matching to tempo.

Converting ReCycle files to Apple Loops

Logic can import and work with Propellerhead's ReCycle files and these can be made into Apple Loops for use across a wide variety of projects. To import ReCycle files into the project choose File > Import Audio File and select the ReCycle file, or drag it to the Arrange area. Logic will next flag up a dialogue box which lets you decide how you wish to process this file. In this instance, you should choose to Render Into Apple Loop. The ReCycle file will now be an Apple Loop in the Audio Bin.

Walkthrough 2

Editing Across Multiple Drum Tracks

Step 1:

With the tracks now grouped, we can now turn to the process of managing the drum parts in our project. We can easily top and tail the grouped drum regions using the Resize tool. In this example here, the drummer carried on playing during a quieter passage of the song where drums were not required. Simply select a conventional tool using the Tool menu (Escape) such as the Scissors for splitting the region into two separates.

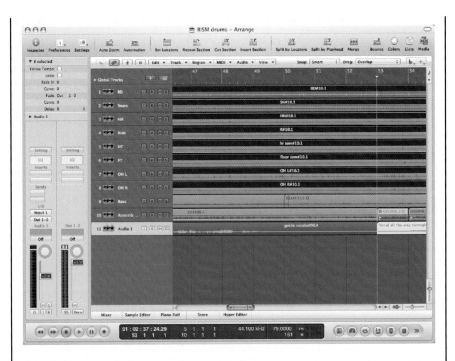

It is worth remembering that the Marquee tool can also be used for editing elements of the regions.

Step 2:

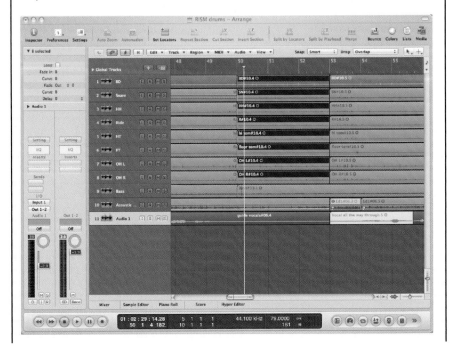

Now turning your attention to the arrangement itself. Select the regions you wish to edit, in this case we want to take out eight bars worth of drums and have selected them. We can either separate the regions by using the Scissors Tool (Escape > 5) cutting at the beginning and at the end of the required section. Or we could draw the Marquee tool around the required section then click once with the Scissors Tool.

Step 3:

To tidy up this part we would need to select the regions and then either mute the regions by pressing "M" or remove them altogether by pressing Backspace. Next, locate a crash hit from the end of the drum part which can be copied to the start of this quieter passage in the song. Use the crossfade tool from the Click Tools menu (Escape + 0) to smooth over the transitions between the two regions.

MIDI sequencing and instrument plug-ins

6.1 Introduction

Besides being a tremendously competent and versatile audio sequencer, Logic has been famed for its abilities at recording and manipulating MIDI data, albeit in the virtual domain – using Logic's own virtual instruments and other third-party plug-ins – or in the now-dwindling world of hardware synths and samplers. Working with MIDI in Logic offers some of the most creative and flexible musical production tools available, offering complete control of every last musical detail in your track: from the precise time and placement of notes, to the exact timbre and sound parts are played with.

In this chapter, we're going to explore some of the key techniques of working with MIDI, from basic region manipulation on the arrange page to more in-depth editing techniques using Logic's various MIDI editors: the Piano Roll, Hyper Editing and the Event List. We'll also explore some of Logic's key virtual instruments – including the EXS24, the CPU-light plug-ins ES1, ES E, ES M and ES P, alongside the EVP88, EVB3 and EVD6 vintage instruments – all tools that you'll undoubtedly use on a daily basis. If you're particularly interested in the creative application of MIDI, and the techniques of sound design, be sure to read Chapter 7, which follows on from many of the concepts dealt here.

6.2 MIDI concepts

MIDI recording in Logic involves working either with virtual instruments embedded within the application itself (like the ES2 synthesizer, or the EVP88, a modelled Electric Piano), or MIDI equipment, like hardware samplers and synthesizers, externally connected to Logic via an appropriate MIDI interface. Unlike Audio recordings, a MIDI recording makes an important detachment between musical data (in other words, the raw note information that forms a musical performance) and patch data (the information required to generate the sounds). Fundamentally, what this means is that within MIDI sequencer we have complete and independent control over each of these elements – we can change the sound, or the performance in subtle or radical ways at any point in the production process.

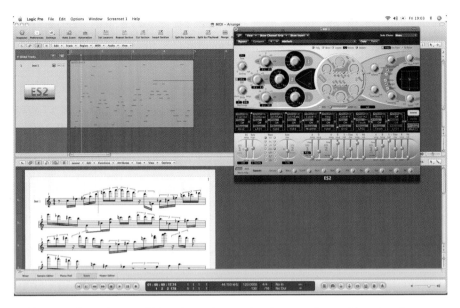

Unlike an audio region, a MIDI region only contains the note data, with a virtual instrument (or MIDI synthesizer) used to control the precise properties of the sound.

6.3 Creating instrument tracks

Creating a suitable MIDI track is done via the New Tracks command (as we saw with the creation of an audio track), although in this example, we have the choice between a Software Instrument (which would be used to access any of Logic's own virtual instruments or third-party audio units like Native Instruments' Absynth or Kontakt) and an External MIDI track, in the example of accessing legacy MIDI hardware.

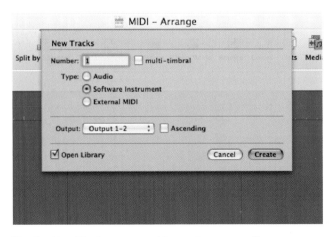

Use the New Tracks command to create a new software instrument track – for either Logic's own virtual instruments or third-party audio unit plug-ins like Kontakt 3.

Knowledgebase 1

What is MIDI?

MIDI was originally developed as a communication protocol between hardware synthesizers, samplers, drum machines, and of course, sequencers like the original MIDI-only version of Logic. The protocol is made up of a number of different types of messages – including note on, note off, velocity, modulation, and so on – all of which can be sent to 16 available MIDI channels. In theory, the channel system was originally used to denote different devices along a "daisy chain" of MIDI leads, with each device sifting its appropriate collection of message from the collective stream of MIDI data passing along.

Nowadays, of course, hardware MIDI devices are much less common (although you may well use a MIDI controller keyboard plugged into the MIDI port of the Audio Interface), but the MIDI protocol is still alive an effective means of "internally" communicating with virtual instruments in Logic's audio engine. Open up the Event list, for example, and you still get to see the list of familiar MIDI events, all with their associated channel and parameter numbers. Also, the Instrument Parameters in the inspector allow you to specify a MIDI channel for the instrument in question (although most default to ALL), and of course, instruments like the EVB3 use different MIDI channels to differentiate between the two different keyboard manuals, and the bass pedals, contained within the instrument.

So, even though you may never connect a MIDI lead again in your life, MIDI is definitely here to stay as an integral part of sequencing with Logic.

The options presented for the new track types vary to that of an audio track, with a few less parameters to choose from. As with audio tracks, you can create multiple instances, which is handy if you know you're going to require a number of different virtual instruments in your session (16, for example, should provide an adequate starting point). If you're using third-party software instruments like Kontakt 3 that feature multi-timbral operation (in other words, you can load multiple sounds into the instrument), you might want to select the multi-timbral option. This will create 16 different tracks, all of which are assigned to the same instrument and will allow you to address the 16 different channels, or instrument slots, within it.

6.4 Instantiating virtual instruments and the Library feature

Assigning a virtual instrument to a corresponding Software Instrument track is done in one of two ways – either selecting from a wide range of so-called Channel Strip Settings, or selecting a specific software instrument that interests you.

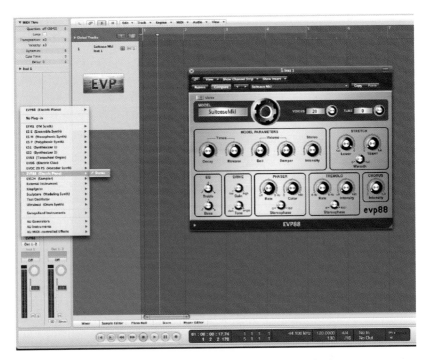

Parameters and presets can be both directly controlled from the instrument plug-in's interface.

Let's start by taking a look at loading in case of a specific software instrument – albeit one from Logic's integral instrument collection, or a third-party audio unit. Possibly the easiest way of doing this is via the Channel's inspector, using the small mixer strip that appears in the bottom left-hand corner or the arrange window. By clicking just below the I/O label (in what should currently be an empty greyed-out box), you'll open the list of Logic's integral software instruments, and, at the bottom of the list, an option to browse AU instruments, which should have all your third-party audio unit instruments organized by manufacturer.

Once the instrument has been instantiated, its interface will be presented to you on the screen. On the top of the interface, you should find a drop-down list of the various presets associated with the instrument, as well as two arrows to sequentially move backwards and forwards through the list.

Browsing instrument settings in the Library

As we've seen in previous chapter, the Media area can be used to browse a list of audio files in our session, the Apple Loop collection, and indeed, your computer's hard drive. Another extension of this is the Library feature, which offers an alternative, list-based way of browsing and navigating an instrument's settings. To access the Library, open the Media area (B) and click on the Library Tab. You should now see a list of your presets settings, possibly organized into folders, through which you can navigate as you see fit.

The library provides a more elegant and seamless way of stepping through an instrument's settings.

Plug-in focus 1

EVP88

The EVP88 was one of the first virtual instruments developed by Logic, modelled on the classic electric piano sounds – like the Fender Rhodes and Wurlitzer – used throughout the 1960s and 1970s. Of all the vintage instrument plug-ins included in Logic, the interface is one of the easiest to comprehend, with a relatively simple set of parameters to work with. The most important of these is the Model parameter, which selects a basic model from which to work. Models range from the classic Rhodes designs, like the Suitcase Mk1 and Suitcase V2, to Wurlitzers (great on rock tracks) and a Hohner Electra.

The parameters found towards the bottom of the interface replicate the more traditional controls and features found at an electric piano player's hands. EQ and Drive, for example, govern the basic tone of the patch – keep these relatively flat if you want a clean DI'd sound, or ramp them up if you need to produce the more coloured tone, like the original Rhodes fed through a fender twin amp. The tremolo is also another feature integral to the sound of the original pianos, with controls to change the rate and intensity of the modulation. Try experimenting with the stereo phase – the 180° setting produces a pleasing stereo autopan, while setting progressively nearer the 0° setting produce an increasingly mono tremolo. The Chorus and Phaser are also important Rhodes-orientated effects, used to great effect in the late 1970s on tracks like 10cc's *I'm Not In Love.*

Experimenting with the model parameters allows you to twist certain "mechanical" properties of the original instruments – variations that might have occurred with the condition of the instrument, for example, or in the way it was manufactured. For example, increasing the Bell volume makes the sound closer to the bright "Dyno Rhodes" modifications performed on Fender Rhodes pianos in the early 1980s.

The EVP88 remains one of the silkiest Electric Piano emulations available, with a surprisingly low CPU drain.

Using Channel Strip Settings with instruments

As an alternative to instantiating a single instrument, you can use Logic's Channel Strip Settings feature. A Channel Strip Setting offers the complete range of settings for the current selected Instrument tracks – instantiating an appropriate instrument, loading samples (where appropriate), alongside the complete set of plug-in effects (reverbs, equalization, compression and so on) that are associated with that instrument.

Arguably, what you get with the Channel Strip Settings feature is an output that's closer to a finished sound as it might appear in the mix, although it's worth remembering that these particular settings might not be entirely appropriate to your eventual balance of sounds. Another potential downside of the extra plug-ins is the increased drain on your CPU – for example, it might be more efficient for several instrument to share a reverb plug-in rather than each instrument having its own reverb settings. Of course, there's always the option of disabling plug-ins (by Alt-clicking on them, or activating their bypass control), or even removing them entirely, should you change your mind at a later point.

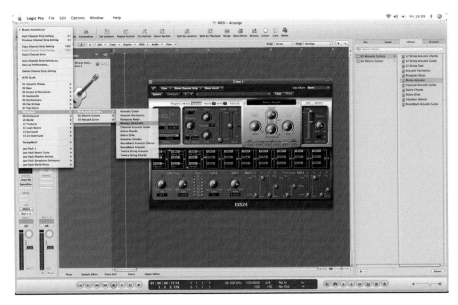

Channel Strip Settings provide a complete combination of instrument and associated effects. This is a useful solution for mix-ready sounds, but the effects might begin to clutter the mix.

The Channel Strip Settings can be accessed via the small grey Setting box at the top of a channel strip. Note that if you've got the Library open, clicking on the same Setting box will also open up the Channel Strip Settings in the Library browser.

Of course, the Channel Strip Settings can also be used if you want to save your own Instrument settings with their associated effects – maybe to move a patch between different project files, for example.

Changing the Instrument Parameters

Instrument Parameters are a global set of parameters associated with a particular instrument. The Instrument Parameters can be found via the inspector, with the current selected track having its parameters displayed towards the middle of the Inspector – remember to open the box (using the small arrow) if it is minimized. Although the Instrument Parameters include some useful features, most users tend to leave these settings in their default positions, unless they're working with external MIDI sources (see Section 6.5 for more information on this).

Changing an Instruments Icon

Icons allow you to easily distinguish between different tracks in your Logic project. By default, any of Logic's instruments will establish a corresponding icon to distinguish between the different models (EXS, ES2 and so on), although you might want to use one of Logic's other icons to define a track's

The Instrument Parameters box allows you to control a number of key parameters in relation to the instrument instantiated into a particular track. On the whole, this can be left in its default state.

musical properties (strings, bass, guitar, drums, and so on). To change the icon, simply click on the icon setting in the Instrument Parameters box.

Transposition and velocity

Use the transposition and velocity parameters to quickly change the way an instrument responds to the performance data it's presented with. For example, if you know you want the instrument to take a prominent, aggressive position in the mix, it might be appropriate to use an additive value (like +20) for the velocity setting. All performance data sent to this instrument – whether played live from MIDI keyboard, or from any track on the arrange window assigned to this instrument – will be modified accordingly. If you want more control over how and when these modifications occur, consider applying them as MIDI edits – either by modifying sequence parameters or one of the MIDI editor windows, both of which we'll see more of later on in this chapter.

Key Limit

This feature is useful where you're orchestrating for acoustic instruments with a known limit in their playable range. For example, a 1st Violin could have

Using icons will allow you to better distinguish between different tracks and their various musical roles.

its lower range set to G2 to remind you not to play or record any MIDI data below this point. Although some of the sample instruments included with the EXS24, or as third-party data, might account for the natural instrument's playable range, not all do. A little forethought here, therefore, could save hours of extra editing and arrangement work later on.

6.5 Working with external MIDI instruments

First off, if you haven't got any external MIDI equipment, you can skip this section and carry onto the next! However, if you're lucky enough to have some external MIDI sound sources, like (proper) vintage synthesizers and samplers, then you'll need to spend a few minutes adjusting a few settings to get things working correctly. It could also be the case that you're also using the MIDI protocol to communicate with other computers in your studio,

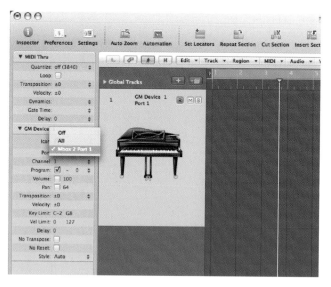

Change an External MIDI Instrument's Port setting (as part of the Instrument Parameters) to assign it to a physical connection on your MIDI/audio interface.

even using a network connection to connect to something like GigaStudio, in which case you'll also need to follow these basic configuration steps.

Because we're working with an external MIDI device, we'll need to ensure that our track is being directed to the right physical MIDI output on our corresponding MIDI or audio interface. Start by selecting the External MIDI track in question, opening the inspector, and looking at the Instrument Parameters. You'll need to establish a Port setting to correspond to the port your MIDI device is plugged into, as well as setting a corresponding MIDI channel. If the instrument is multi-timbral – in other words, a device that allows a different sound to be placed on each of the 16 available MIDI channels – you might want to run multiple External MIDI tracks, each set to a different MIDI channel. To do this, go to the Track menu and select New with Next MIDI channel (or Apple + Alt + M). You can also control-click at any point on the track name to re-assign the object to another MIDI channel accordingly.

Plug-in focus 2

EVB3

The EVB3 is Logic's own virtual recreation of the classic Hammond Organ. The original instrument used a unique tonewheel sound generation system, controlled by a series of drawbars, to create a surprisingly rich and varied set of organ sounds. The package was then made complete by the addition of a rotating speaker cabinet – otherwise know as the Leslie – providing a distinctive swirling movement to any sound that passed through it.

Like the original B3, the EVB3 supports two manuals (or keyboards) and a set of bass pedals, each with its own set of drawbars. To access these, you'll need to set the MIDI channel in the instrument's parameter box to ALL – this way the EVB3 will be able to receive MIDI data on all channels, with Channel 1 being directed to the upper manual, Channel 2 to the lower manual, and Channel 3 to the bass pedals. Alternatively, change the Keyboard Mode from Split – you should now find the various manuals accessible from different octaves of your controller keyboard.

Shaping the tone of the EVB3 is achieved by sliding the relative position of each manual's drawbars. Essentially, the drawbars represent a different harmonic from the harmonic series – try using a simple combination of just a few of the lower drawbars for a mellow, pure tone, or "pull out all the stops" for a fuller timbre that is rich in harmonics. The colours also provide some indication to their respective roles: the white drawbars being even harmonics, black drawbars containing the odd harmonics, and the brown drawbars representing the sub (an octave below the fundamental) and third harmonic.

Switching between the different Leslie settings (Chorale, Brake and Tremelo) forms a distinctive part of the movement and progression in a Hammond performance. The Chorale creates a slow vibrato, the Tremolo a fast-moving vibrato, whereas brake stops the vibrato effect. As well as being able to switch the settings from the interface (these could, for example, be recorded into MIDI), you can also use the mod wheel to move through the three settings.

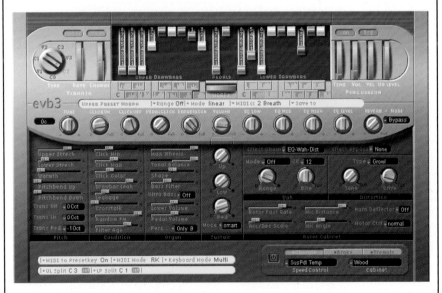

The EVB3 is Logic's take on the classic Hammond organ, with two manuals and set of bass pedals, and of course, the authentic drawbar sound creation system.

How do I monitor MIDI sound sources?

Although you may have routed MIDI out of Logic to the device in question, there's still no guarantee that you'll be able to monitor it correctly. Unlike Audio Instruments, of course, an external MIDI instrument won't have a corresponding mixer channel to monitor its output (although you do have a fader to remotely control the level and pan settings via MIDI). One solution is to run your main outputs from Logic into a physical mixer (Channels 1 and 2), with your synthesizer, sampler or drum machine connected into any remaining spare channel. With this setup, however, you won't be able to use any of Logic's own effects on the synthesizer's output, or indeed, use the bounce to disk mixdown functionality to create a complete version of the mix.

A more appropriate solution is to plug the outputs of your MIDI hardware sound sources into spare inputs on your audio interface. Next create a number of spare Audio Tracks (one for each device connected to your audio interface), with each of the audio tracks set to Input Monitor. Return back to your External MIDI track, press a few keys on your controller keyboard,

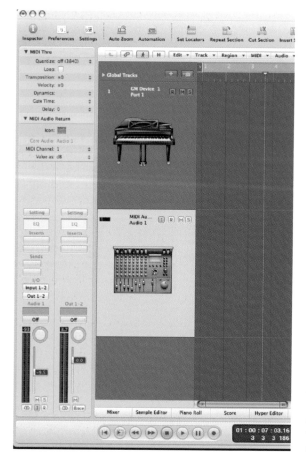

To monitor a MIDI synthesizer, you'll need to connect it to your audio interface's audio inputs, create a corresponding audio track and place this into Input Monitor mode. This will also eventually allow you "print" the synthesizer's output as an audio file.

and you should now hear the sound source being routed back into Logic's main mix. Note that you might need to adjust the input gain on your audio interface to account for the levels being sent out from your sound source, and that the monitoring will, of course, be subject to any latency present in your system.

The significant advantage of your MIDI sound module being routed in this way is that you have complete control over how it appears in the Logic mix, even with the possibility of adding plug-ins, like equalizers, compression, and reverb. Of course, if you're completely happy with the performance and sound, you can quickly flick from Input Monitor to Record and have the part permanently stored as an audio file, which might offer a better long-term solution as well as being a convenient means of negating any latency issues.

What if I've got more than one external MIDI device?

Adding further MIDI devices beyond the basic GM Device included in the basic Empty Project file goes beyond the scope of the main arrange window interface. Instead, you'll need to create new instrument objects as part of the environment, which is something we'll cover in more detail in Chapter 11. If you have a particularly complicated MIDI setup, you might want to consider building your own Template file so that all your pre-existing MIDI devices are automatically set up with each new project. Again, this is something we'll specifically look at in Chapter 11.

My MIDI tracks are playing back slightly late, how do I rectify this?

As we mentioned before, it is feasible that the monitoring of external MIDI tracks via your MIDI hardware could be causing some minor latency issues, with the MIDI tracks appearing to play slightly behind the beat. To rectify this, you might want to consider applying a negative delay via the Instrument Parameters so as to send the MIDI data out slightly ahead of time and there-fore negate any of the latency involved. Finding the right settings (which will probably be in the region of about −20 or so, depending on the latency in your system) could be as simple as adjusting the delay parameter by ear. Alternatively record in a bar or so of MIDI information as an audio file, and then adjust the instrument's negative delay until the "live" MIDI data and recorded MIDI data playback exactly in sync.

One important factor to note here is that once a MIDI track has been recorded as audio, the monitoring latency will be removed. It is important, therefore, that if you intend to record an external MIDI part into Logic, that you ensure that the delay parameter has been reset to its zero point.

To throw yet another spanner in the works, it should also be pointed out that Logic's plug-in delay compensation (found under Preferences > Audio > General), where used with latency-including DSP-acceleration systems like Universal Audio's UAD-1 or TC Electronics' PowerCore, can also create

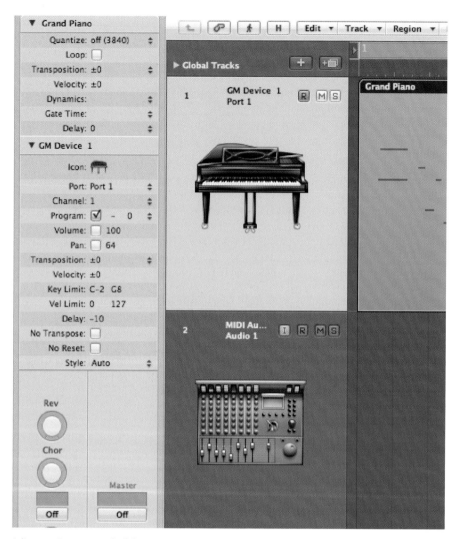

Adjust an Instrument's delay settings to account for any monitoring latency.

additional MIDI playback latencies that will vary given the amount of plug-ins, and where they are placed in Logic's mixer. The best advice here is to avoid applying these types of plug-ins until the mix, or use the low-latency monitoring mode to temporarily bypass the offending plug-ins while you're are monitoring External MIDI instruments.

6.6 Making a MIDI recording

Making a MIDI recording follows many of the same procedures explored in Chapter 4, although you might want to note the few dissimilarities.

Where's the data?

The MIDI data created with a recording stays within the region, rather than referencing a separate audio file, as is the case with an audio region. MIDI data takes up an incredibly small amount of memory, with a whole project's MIDI resources utilizing less than about 1 MB of RAM in total.

Recording in cycle mode and overdubs

Unlike an audio recording, which will create a take folder for parts recorded in cycle mode, a MIDI recording will simply layer addition passes into the same region. The assumption with cycle mode is that you use the repeating playback to record each part a layer at a time (like a kick, followed by the snare), without stopping the transport at any point. If you want to record different takes of a keyboard solo, for example, consider creating a new track assigned to the same instrument (Track > New with Same Channel Strip/Instrument, or Alt + Apple + S), with previous take muted. Recording directly over the first take (as you would with an audio overdub) will simply create regions placed on top of one another, all of which will play at the same time. Multiple stacked MIDI regions – unlike take folders – can get difficult to manage, so try to keep on top of any layered recordings using the editing and arrangement techniques described in the following parts of this chapter.

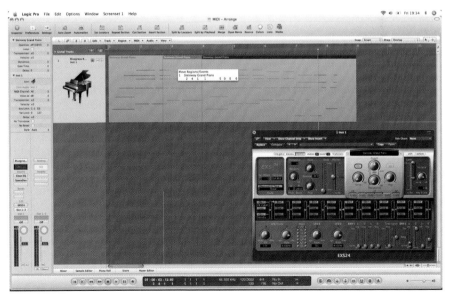

Multiple MIDI recordings will simply create "layered" takes, which can get confusing on a single MIDI track. Consider creating duplicate instrument tracks to better distinguish between your various takes.

In this example, a duplicate track has been used to record the second take of a MIDI performance, rather than have it layered onto the original MIDI region.

Plug·in focus 3

EVD6

The EVD6 is Logic's virtual recreation of Hohner's "funkalicious" Clavinet. The original instrument worked like an electrified harpsichord, with a number of reeds plucking a series of strings, with the resultant vibrations being picked up by a collection of guitar-like pickups mounted in the unit.

As with the EVP88, the EVD6 divides its interface up into areas with more conventional controls – as found on the original instrument – as well a detailed set of modelling parameters to adjust every last detail of the EVD6's output. The conventional controls focus on the rocker switches towards the bottom of the interface. These move between basic, preset filter positions (Brilliant, Treble, Medium, and Soft) and different pickup combinations. For example, by alternating the various filter rockers, you can move between a soft, almost muted tone, to a full and raspy clavinet pluck. The pickup switches produce more subtle transformation in the sound, and are worth experimenting with to extract just the right flavour from the instrument.

Along the top of the interface are the various modelling parameters – including some interesting options to physically move the pickup positions inside the virtual clavinet – and the various additional FX options. In keeping with the approach of many keyboard players in the 1960s and

1970s, the EVD6 includes a Distortion section, modulation effects (including phaser, flanger and chorus), and the infamous Wah-wah. Although the Wah-wah can be set to track the sound's envelope (for an auto-wah effect), the best solution is to map it to something like the modulation wheel, or a spare MIDI controller, to keep it as hands-on as possible.

Logic's EVD6 is a faithful recreation of Hohner's "funkalicious" Clavinet – complete with the original controls and the number of modelling options.

6.7 Editing and arranging MIDI regions

Again, the process of editing and arranging MIDI regions on the arrange window follows many of the same conventions as audio editing, allowing you to sketch out and develop your music arrangement over time. As we saw in the audio editing chapter, regions can be resized, cut, moved and duplicated using the same tools and processes involved in audio editing. However, as with the process of recording, it's worth pointing out a few differences with respect to working with MIDI regions.

Merging MIDI regions

Using the glue tool allows you to merge a number of MIDI regions into a single contiguous region, although unlike audio data, this doesn't result in the creation of a newly rendered audio file. Merging MIDI regions is a great way of combining any stacked MIDI regions (where regions have been stacked on top of one another), allowing you to see all the MIDI data in a unified way.

Working without fades and crossfades

As you'd expect, a MIDI region doesn't require any crossfades to be placed at edit points; therefore, it is impossible to apply fade-based processing to a MIDI region. To achieve the same volume contouring, you'll need to turn to either Automation (for more information see Section 8.10 in Chapter 8) or the Hyper Draw feature (covered in Section 6.17).

Timestretching MIDI regions

Although it is possible to perform a certainly level of time adjustment with audio regions, it is MIDI regions that offer some of the greatest flexibilities with respect to time-based properties. By Alt-dragging the end point of a MIDI region, for example, you can quickly time compress or expand the data that are contained within it. This can be an interesting way of exploring double-time or half-time treatments by stretching the region by 200 or 50% of its original size. Of course, as this is simply the playback of MIDI data you're effecting, there is no loss to the sound quality of playback, which would easily become apparent with the same level of stretch applied to an audio file.

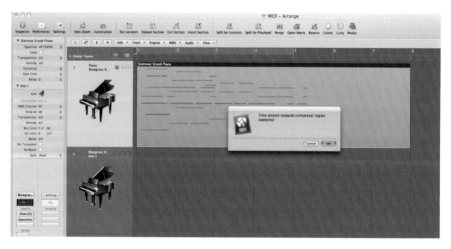

Using the Alt key as you resize a MIDI region will allow you to time compress or expand its MIDI data accordingly.

It should also go without saying that a MIDI region follows and responds to the project's tempo, without needing to activate the Follow Tempo feature in the region parameters box. This also makes it feasible to record a MIDI region at more comfortable playing speed, and then increase Logic's tempo to playback the region at the appropriate speed. The only issue to be aware of here, though, are accompanying audio tracks, which won't necessarily follow the same tempo change. In this case, simply monitor from virtual instrument or MIDI tracks accordingly.

6.8 Region parameters: quantizing and beyond

Once you've recorded your first few bars of MIDI data, you can now start to explore some quick-and-easy modifications directly from the arrange window. One of the Logic's key concepts is that a large number of MIDI modifications can be performed directly from the arrange window, without ever having to go anywhere near a conventional MIDI editor (like the Piano Roll or List editor windows). To perform these modifications, simply select one or more regions and turn your attention to the region parameters box in the top left-hand corner of the arrange window (if you can't see the sequence parameters, then make sure the menu option View > Extended Region Parameters is engaged).

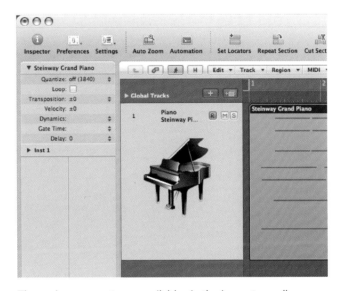

The region parameters – available via the inspector – allow you to quickly define key properties of your MIDI regions.

The main parameters accessible from the region parameters box include the region's name, quantizing, loop, transposition, velocity, dynamics, gate times and delay settings. What the region parameters allow you to do is to quickly define some important overriding concepts of how the region – or group of regions – is played back. No data are modified – these parameters are simply "playback" criteria. For example, you could select the entirety of your MIDI composition and adjust the transpose option to find a key that best suits your vocalist's range. Should you decide to return your composition to its original pitch, simply select the regions and reset the transposition accordingly.

Arguably, the most important function of the region parameters is the Quantize tool, allowing you to quickly tighten the rhythmic qualities of a given performance. Clicking towards the right-hand side of the sequence

145

parameters box allows you to open up a list of the Logic's available quantize options, ranging from various 1/8, 1/12, and 1/16 note settings, to Swing quantizing, and combined 16/24 options. On the whole, most conventional musical applications will adopt a simple 1/16th note quantizing, so that any stray notes are pulled to a rigid 16th (or semiquaver, to give it its musical name) grid. A swing quantize, on the other hand, shifts every other 16th, producing a distinctive shuffle effect commonly used in Hip-Hop or R n B productions. In swing quantize, the various letters (16A, 16B, 16C and so on) denote the relative amount of shuffle, with 16A, for example, delivering a lighter shuffle than 16D.

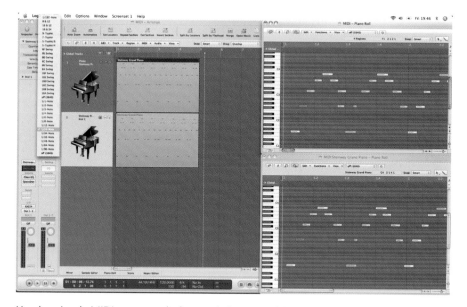

Here's a simple MIDI sequence before and after quantizing. Note how the timing of the quantized region (top) conforms exactly to the project's tempo grid.

Loop – as with the same function in an audio region's region parameters – can set an indefinite loop of the MIDI region until either the end of the project, or another sequence object, is encountered. Transpose, of course, deals with the relative pitch of the MIDI sequence, allowing you to transpose a region, or group of regions, up or down by a selected amount of semitones.

Velocity and Dynamics allow you to control the relative ferocity that a part is played at – setting a number in the Velocity parameter, for example, will either add or subtract a value to the current velocity levels contained within the part. The Dynamics parameter is even more interesting, in that it works like a MIDI compressor or expander – reducing or expanding the range of velocity levels that occur in the region. Indeed, couple these functions with an expressive virtual instrument, and you can have a surprising impact on how the instrument sits in your mix.

Gate Time deals with the duration of notes – changing an elongated series of notes, for example, into a shorter staccato effect. You can achieve some interesting effects here on repetitive sequencer lines, especially where all notes are of fixed duration. Try experimenting with the different gate times, and seeing how this has an effect on the finished output.

Plug-in focus 4

ES E, ES P and ES M

Although it's easy to get carried away with many of Logic's deeper and more complex software instruments, there's still a lot of use to be extracted from some of its simpler synthesizers. One of the best qualities of these instruments are their simplified user interfaces, often with a bias towards specific sound types like pads; in the case of the ES E, or leads and basses; and on the ES M.

Arguably the simplest synthesizer in Logic is the ES E, which is dedicated to simple pads sounds, all from a single oscillator. Try experimenting with some of the key parameters (in other words, the ones with the bigger knobs), exploring the different waves on offer, and tools like cutoff and resonance. The three modes of chorus (Chorus I, Chorus II and Ensemble) are vitally important in creating warmth to the pad sound, with each setting offering a slightly different flavour.

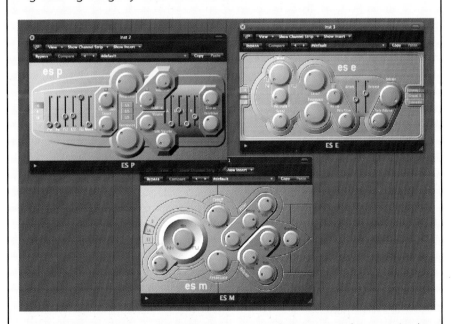

Logic's CPU-light collection of synths can be a great place to turn to for some simple keyboard sounds: like 1-oscillator synth basses or sawtooth pads. Note how each synth is dedicated to a certain range of sounds, making the interface un-cluttered and easy-to-use.

ES M is a virtual instrument modelled on the simple synthesis capabilities of early monophonic synthesizers, like the Roland TB-303. This is a great source for "acidic" basslines, full of distortion and resonance, as well a simple lead synthesizer parts. The Glide parameter is an interesting feature – and a quint-essential part of the aforementioned TB-303 bass sound – allowing the ES M to gliss from one note to the next whenever the first note is held over.

The ES P offers the most complex set of sound possibilities of the three synths, and can create a range of different synth sounds. One of the most distinctive features is the waveform mixer, which allows you to blend a number of different waveforms to create a starting point for the rest of the synth. The first three sliders govern basic waveform types (triangle, sawtooth and square), while two of the final three sliders add square waves an octave below, and two octaves below that of the fundamental pitch. Further sound-shaping features include envelope generators, chorus and an overdrive section.

6.9 The MIDI Thru function

If no region is selected in the arrange area, you should notice that the region parameters box displays MIDI Thru rather than an appropriate region name. As innocuous as the MIDI Thru function may seem at first, it is, in fact, a particularly powerful feature of sequencing in Logic. MIDI Thru's most immediate and noticeable benefit is as "live" modifier of MIDI data coming into Logic. For example, rather than playing in a difficult key, like C# major, you could choose to play in C major and use the transposition parameter to shift everything that you play up half a semitone. Equally, the Velocity parameter can be a useful way of increasing or decreasing the relative loudness as you play.

Another important function of the MIDI Thru option should also become evident as you make recordings into Logic. In effect, the MIDI Thru function sets a default region parameter setting for everything you record into Logic. Going

Using the MIDI Thru option as part of the region parameters can preset certain playback criteria – like quantizing or transposition, for example.

back to the transposition example, you should note that a part recorded in with MIDI Thru set to +1 transposition will have its region parameter set accordingly (this also means you can go back to the data in its original form). This MIDI Thru functionality is particularly powerful with a setting like Quantize, where once set, all parts will be automatically quantized on-the-fly, immediately after they've been recorded.

6.10 Extended region parameters

Just to illustrate just how far you can edit a MIDI region before going anywhere near an editor, we can also explore further modifications using the extended sequence parameters. To see the extended set of region parameters, select View > Extended Region Parameters from the arrange window. On the whole, this particular set of region parameters offers an expanded set of quantizing options – particularly with respect to softer, or more refined, quantize settings.

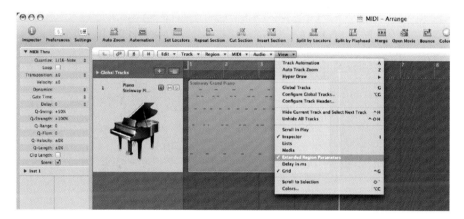

The extended region parameters offer more precise control over the groove and strength of quantize applied. Use these where you need to preserve important qualities of the original performance.

Arguably one of the most important options in the extended region parameters is the Q-Strength parameter. As the name suggests, this directly modifies the "gravitational" strength of quantizing, so that a 100% setting, for example, pulls the notes completely to their grid position, while a setting of 50% only brings the notes halfway. This is a great way of achieving a soft quantizing – improving the overall feel of the performance, without turning it into a metronomic sequence.

Q-Swing offers a finer control over the swing setting than the 16A, 16B, 16C and so on, settings in the standard quantize menu. At 50% the feel is straight, while settings either way (60 or 40%) imparting different qualities of shuffle. If you're particularly into the groove of a piece of music, then the precise control offered here will be important in creating the right feel between parts.

Q-Flam avoids the sometimes-unnecessary side effect of a quantized chord having its notes aggressively bunched together. By using the Flam setting you can "fan-out" the notes, with positive settings producing an upward movement, and negative settings producing a downward movement. Although this won't be applicable to every instrument you work with, the Flam option is a feature that works really well with harpsichord or acoustic guitars samples.

One of the more difficult tools in the quantize tools set to use is the Q-Range control. Q-range is another form of soft quantizing, although in this case, the notion is to ignore notes closer to or on the beat, and instead direct the quantizing engine towards notes that stray beyond a given point. On a setting of zero, for example, all notes will be corrected to the current quantize settings. By decreasing the range into its negative setting, only notes that fall outside the range will be quantized, effectively creating an increasingly soft form of quantizing.

6.11 Normalizing sequence parameters

Should you reach a point where you're completely happy with your region's parameters, you might want to consider making the "adaptive" edits permanent – a process know as Normalizing. Although you'll lose the possibility of going back to the original performance, a normalized region might be easier to handle in the long run, especially if you mange to undo your perfected edits by mistake.

The appropriate functions can be found under the menu option MIDI > Region Parameters, with various options to Normalize region parameters, or destructively apply any Quantize settings you've created.

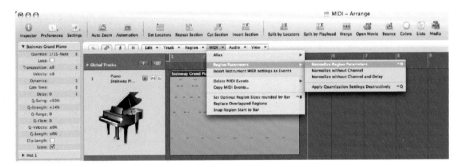

Use the Normalize feature to permanently "fix" your region parameters like quantizing.

6.12 MIDI editing in Logic

Although there's a surprising amount that can be achieved using just the sequence parameters alone, eventually you hit a point where you need to explore a more detailed set of editing options.

Editors can be opened in a number of ways. For example, double-clicking on a region will open the default assigned editor (Preferences > Global, under the Editing tab), which is initially set to the Score editor, although most users will change this to the Piano Roll editor. Alternatively, once one or more regions are selected in the arrange window, you can also use the Piano Roll, Score or Hyper editor tabs at the bottom of the screen to open the respective editors. Finally, using the window menu, you can also open separate editing windows away from your main arrange, which is a useful solution if you're using multiple displays.

Delving deeper still, it's also possible to view and edit your MIDI data in the form of a list of Event data – in other words, the precise note on, note off, and controller movements that your sequence is comprised of. Although this might seem confusing at first, it does offer a precise and informative view of your MIDI data (much the same way a Logic sees it), allowing you to position MIDI notes on a tick-by-tick basis, or set controllers to exact parameter levels.

Ultimately, good MIDI editing practice in Logic is about matching the right MIDI editor with your precise type of modification you want to achieve. For example, if your edits are generally musical, look towards the Piano Roll or Score editors. If your edits involve a more intricate manipulation of control data, then either the Hyper editor or List editor might be more appropriate. Not only will the right editor allow you to work quicker, but you might also discover some interesting creative possibilities along the way.

6.13 The Piano Roll

The Piano Roll editor provides one of the most intuitive and informative ways of editing MIDI data in Logic and is often the first choice for the majority of

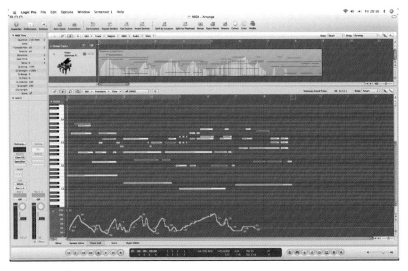

Here's a typical MIDI region being edited in the Piano Roll editor, with note data towards the top of the interface and controller data towards the bottom.

MIDI editing tasks. The editor can display both MIDI note data in the top half of the interface, with the added option of viewing controller data, like pitch-bend and modulation towards the bottom of the editor. Notes are displayed in much the same way as regions in the arrangement page – with the horizontal axis indicating their timing and length, while the vertical axis indicates their relative pitch. In addition to the pitch, timing and length of the notes, you can also see their velocity, displayed as a series of colours.

Plug-in focus 5

EXS24 Main Interface

As a completely integrated solution for using samples, Logic's EXS24 is one of its most powerful music production tools. Compared to other software samplers, like Native Instrument's Kontakt 3, the EXS24 appears a much more approachable tool, although this is largely because it makes a distinction between the front-end controls (which we're looking at here) and the behind-the-scene mapping of samples (which we'll explore in more detail in Chapter 7).

The interface is divided into a number of key sections. Most important is the Sample Instrument menu, with the current selected instrument displayed in the top part of the ESX24's interface. Clicking on this will call up the list of EXS24 Sample Instrument stored on your Mac, as well as Sample Instrument that you might have specifically created for the project in question. You can also browse through, instrument by instrument, using the small + and − buttons either side of the Sampler Instru-ment menu.

Once a sound has been loaded in, you can then start to experiment with the various parameters settings on the interface. These include basic instrument tools – like volume, tuning, total available voices, and so on – as well as a number of synthesis-orientated features, like a multimode filter

Unlike other third-party samplers, EXS24 is specifically designed so that you load multiple instances – one EXS24 for each sound, or instrument, you want to appear in the mix.

(just below the Sampler Instrument menu), three LFOs, two envelope generators and a modulation matrix. In theory, taking the basic mapping of the samples, you can use the range of synthesis tools to transform the basic qualities of your sound, making its timbre darker, for example, or using the envelope to change the attack characteristics. We'll be taking a more detailed look at these possibilities in Chapter 7.

Unlike certain other third-party software samplers, the EXS24 isn't designed to work with multiple instruments in the same instance. Being directly part of Logic's code, the EXS24 is extremely CPU-efficient, with the intention that you run multiple instances for each sample instrument you require in the mix.

6.14 Typical editing scenarios in the Piano Roll

If you're used to editing regions in the arrange window, you should find that many of the techniques carry directly over to the editing in the Piano Roll editor. As with the arrangement window, this editing process is somewhat tied up with a series of tools, including familiar objects (like the pointer tool, pencil tool, and so on) alongside two tools that are unique to the Piano Roll – the quantize tool and the velocity tool.

To understand how the various tools work and relate to each other, let's look at a variety of editing scenarios and how you might combine the various tools and techniques in the Piano Roll editor to rectify this situation.

Correcting pitch

Using the pointer tool, you can move a note vertically up or down to change its relative pitch. By adding in a few keyboard modifiers, you can also

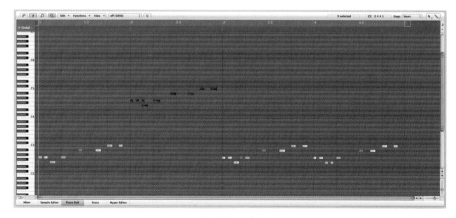

Use Alt + Shift and either of the up or down keys to move selected notes in octaves.

perform some quick modifications without needing to drag objects about. For example, press Alt + either the up or down arrows, and you can move a note or selection of notes up or down in semitone steps. Add in shift (Alt + Shift + Up/Down) and you can move the same notes in octaves.

Correcting timing

Although quantizing can offer a quick-and-easy route to correcting the timing problems of a particular MIDI sequence, sometimes it can be important to have a more precise control over the placement of notes in your sequence. By leaving the region un-quantized, you can use the pointer tool to manually position single notes, or groups of notes, by hand.

As with audio editing, it's important to note the use and operation of the Snap feature (found in the top right-hand corner of the Piano Roll) and how this interacts with the movement of the notes on the Piano Roll's grid. In most cases, this is set to its Smart setting, allowing you to place the notes anywhere along the timeline, although with a definable "pull" to the grid points. If you specifically want the notes to always snap to the grid, you can change this to beat or division, for example, depending on how fine you want the resultant grid. For truly precise movements, though, it's worth adding in a few keyboard modifiers. Shift + Ctrl will switch off any grid setting – irrespective of the current snap setting – allowing you to precisely move the note in question. Alternatively, use the nudge feature (Alt + the left or right arrow) to move the note tick-by-tick using the keyboard.

If you want to get creative, it's also worth experimenting with the timing of phrases by shifting them about in various divisions. For example, try taking an existing phrase and moving this by 1/16th, 2/16ths, or 3/16ths. Notice how the feel of the phrase has changed with the notes falling in a different placement. You can also use this to good effect by starting a track with a phrase that doesn't fall on the first downbeat of the bar, then, as the drums come in, the listener experiences mental "shift" as they're forced to relocate the bar.

As you'd expect, there's also a sophisticated range of quantizing tools available from within the Piano Roll editor. This will be discussed later in Section 6.15.

Duplicating a phrase

By adding in the Alt modifier as you drag your selection, you can create a copy of the note(s) in question. This is an effective way of duplicating musical material within the region.

Correcting length issues

The exact length of a note can be as important to the groove of a part as the note's start point. The specific tool assigned to this task is the Finger Tool, although you can achieve a similar result using the pencil tool and pointer

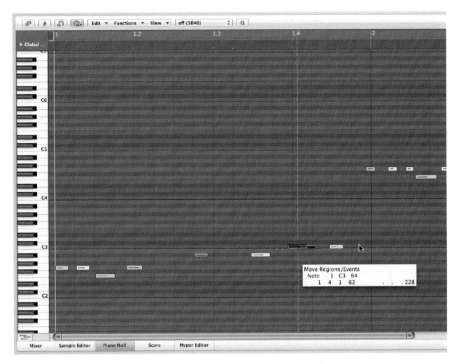

Holding Shift + Ctrl as you move a note allows you to negate any pre-existing snap settings and place each note to a precise tick-based position.

tools by clicking towards the end of a note's duration. Try selecting one or more notes and reducing their length to create a sharp, staccato-like playing style. Alternatively, try lengthening the notes to produce a smoother, more legato-like effect. Mixing and matching these different length settings can be surprisingly effective on basslines.

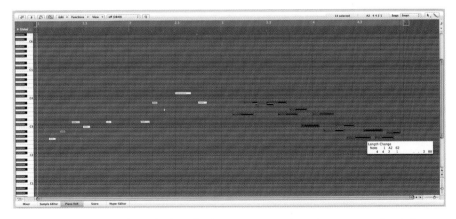

Use the finder tool to adjust the length of one or more selected notes – moving between a staccato and legato playing style.

Changing velocity

Velocity – or in other words, how hard you play a note – has a big part to play in the expression in your music. Ideally, the best types of music incorporate a range of different velocities – both from note to note, and over the whole "emotional arc" of the piece. The velocity tool in the Piano Roll editor allows you to change the velocity of any note or group of notes you click on. Simply click on the note and move the mouse up or down accordingly to increase or decrease the current velocity accordingly. Where a group of notes is selected, the velocity will keep their respective balance (in other words, some notes louder than other).

One aspect of editing where velocity really comes in important is drum programming. As you're putting together a pattern, be sure to experiment with different accents on the various main beats, as well as adding discrete ghost notes using lower velocity settings.

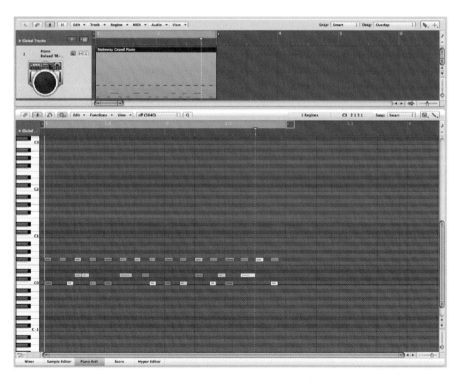

Here's an example of programmed drum loop making extensive use of the velocity tool to define the accent and grove of the loop.

Although the velocity tool provides one solution to changing a note's loudness, it's also possible to use the Hyper Draw controllers display to also achieve the same task. Have a look at Section 6.16 for more information.

Creating new notes

Use the pencil tool as a means of creating new notes in your region. As we saw in the section on changing length issues, the pencil tool can also be used to define the note length by clicking towards the end of the block. One other trick to look out for is how the pencil defaults to creating a note based on the last edited note event. For example, if you create a note and modify to be 1/64th long with a low velocity, all subsequent notes created with the pencil tool will conform to that standard. This can be a useful way of setting a default note type for all subsequent newly created notes to follow.

Knowledgebase 2

EXS24 virtual memory and disk streaming

In an attempt to make a realistic recreation of acoustic instruments, many third-party sample libraries can make quite exhaustive demands on your RAM. For example, some Piano instruments can be comprised of multiple velocity layers, with a separate set of samples for each key on an 88-note piano keyboard. It's not uncommon, therefore, for these instruments to have RAM demands between 1 and 2 GB – something that many user's systems will fail to cope with.

If you do run out of available RAM (the sampler will tell you that it has been exceeded and will fail to achieve a complete load), you might need to consider using the EXS24's Virtual Memory feature. Virtual Memory works by only loading a small amount of the start of each sample, with the remainder streamed from your hard drive as and when required. By using Virtual Memory, you'll be able to load a far greater number of these large instruments into a project and have them played back successfully.

To access the Virtual Memory feature, open up the EXS24 Options menu and select Virtual Memory from the list. A dialogue box will allow you to set and optimize settings, balancing off the hard drive's speed with the amount of disk activity (namely, audio track streaming and recording) also happening at the same time. Obviously, if you're already running a lot of audio tracks, this might begin to put your hard drive under an increasing amount of strain, whereas a low amount of disk activity will ensure opti-mal performance with a large amount of samples voices being playable at the same time. You might also want to consider using an external Fire 800 drive, or an additional SATA drive (internal or external), to keep certain types of data accessible quickly and easily.

In most cases, the usual "Core Audio Overload" will indicate the system being pushed too hard, although you might also want to listen out for lon-ger notes tails, which can get snatched off by excessive disk activity. Also check out the Disk I/O Traffic and Not Reading from Disk in Time performance indicators in the Virtual Memory Dialogue box. These will also indicate the effective, or problematic, use of your system resources.

With Virtual Memory engaged, you'll be able to load a greater number of RAM-hungry software instruments. Be aware, though, of potential knock-on effects on your system's performance.

Deleting/muting notes

To remove a note from a region, use the Delete tool, or simply select the required notes and hit delete on your keyboard. However, a more strategic solution might be to use the mute tool. By muting the note event, it will cease to be played back from the region, but at any point you can return to the region and un-mute the event to hear the original note as it was once played.

One other useful application of the muted notes feature is as an editing guide. For example, say you're trying to harmonize or augment an existing sequence using a different virtual instrument. By copying the region and muting the

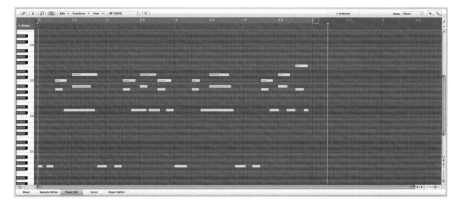

Muting notes is a useful alternative to delete them. In this example, a muted line has been used as a guide to create a harmonized part.

original notes, you can visualize what the other part is playing, while at the same time constructing or editing a new sequence to play over the top.

6.15 Quantizing inside the Piano Roll editor

Despite the Inspector being the first port of call for quantizing tasks in Logic, it's also interesting to note that another layer of quantizing exists within the Piano Roll editor itself. As flexible as this is, that this two-tier system can appear somewhat confusing at first, so it's well worth spending a few minutes familiarizing yourself with the interaction between these two layers of quantizing.

Start by recording a simple sequence into Logic and open this up in the Piano Roll editor. At this point, you should see the various imprecise placements of the notes to match what you originally played. Now move up to the region parameters, as part of the Inspector, and set in Quantize setting to an appropriate value, like 1/16th. Glancing back at the edited part, you should now see the notes snapped exactly to the grid.

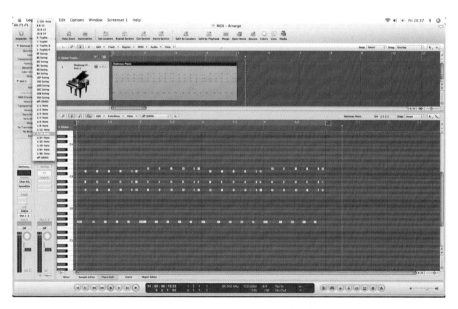

Quantize settings made in the region parameters box will automatically be visualized in the Piano Roll editor.

Now let's try quantizing from directly inside the editor itself. One technique for quantizing is to select one or more notes in the Piano Roll editor and then pick a desirable quantize setting (like 1/16 note) from the drop-down quantize menu in the editor window itself. Another technique is to select the desired quantize setting first, then use the quantize tool to nudge the appropriate

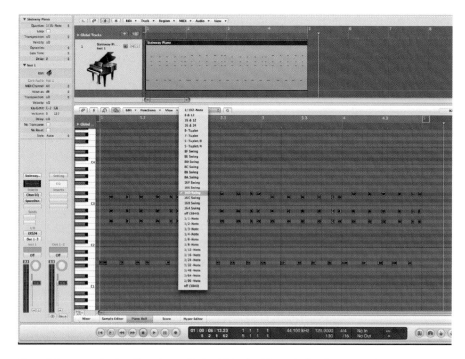

Quantizing can also be directly applied within the Piano Roll editor, negating any exiting quantize settings that have already been set in the region parameters box. This allows you to quantize on a note-by-note basis.

notes into line (this is a good way of being slightly more selective with your quantizing). Note, however, that with both techniques the settings with the Piano Roll editor ignores any settings made in the region parameters. However, should any settings be re-applied in as part of the region parameters, then the current quantize settings (in the Piano Roll) are lost.

6.16 Working with controller data using Hyper Draw

Although the Piano Roll editor does an excellent job with note data, it doesn't quite provide the full picture with respect to MIDI controller data – including dynamic movements of modulation, pitch bend, volume and pan data. However, by opening up the Hyper Draw view at the bottom of the Piano Roll window, you'll be able to see controller data as a series of lines with nodes along their length to edit the respective parameter levels. To open the Hyper Draw view select View > Hyper Draw (followed by the controller of your choice from the list), or open the window using the tab in the bottom left-hand corner of the Piano Roll window.

To better understand what can be achieved in the Piano Roll editor, let's take a look at some typical editing tasks.

Editing Velocity in Hyper Draw

Although we can edit velocity data in the conventional note display of the Piano Roll editor, there are several operations – like a gradual rise or decay of velocity, for example – that can be better achieved using Hyper Draw. To edit velocity using Hyper Draw, you'll need to select the Velocity option from the drop-down list accessible from the small arrow icon to the left of the Hyper Draw area. Once the Velocity view is active, you should be able to adjust the individual velocities, although its most powerful feature is creating a gradual fall or rise over time. To do this, click and hold with the mouse, only releasing when you find a suitable "start point" for the line. Now, move the pointer over to end point, noticing the trail that is left behind it. When you click again, the virtual line will be drawn, effectively raising or lowering the velocities accordingly.

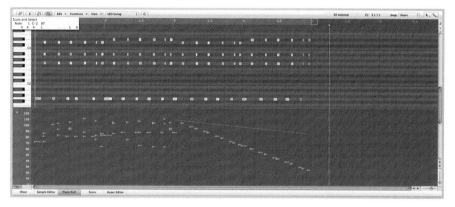

Editing velocity using the hyper view is a much better way of drawing in a gradual rise or fall of velocity.

Editing Volume

Volume moves – like a number of other generic controllers including pan, modulation and pitch bend – can also be drawn directly into the Hyper Draw area, allowing you to creatively shape an instrument's sound proprieties over time. In the specific case of volume and pan (assuming you're using it to control a software instrument), you'll actually see the volume and pan controls on the mixer move to reflect the curves you've drawn in the hyper view.

Drawing curves for these controllers can be done in two ways. The first technique is to use the pencil and effectively draw in the moves "freehand," although, in fact, Logic will actually place a series of node points along the duration of the part. The arrow tool allows you to place, edit or remove the nodes one by one, which might be a better solution for more controlled movements.

Editing other controllers, including filter cutoff

If you can't see the controller that you need in the pop-up list, you can always use the Other option to access any 1 of the 127 different MIDI controller

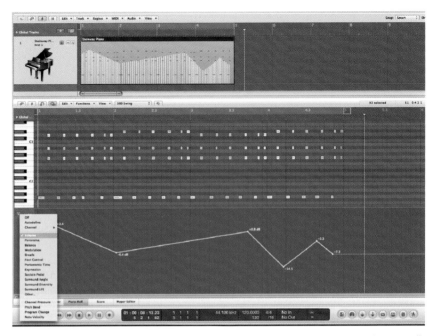

Hyper Draw can also provide an effective solution for drawing in more generic controller moves – like volume or pan, for example.

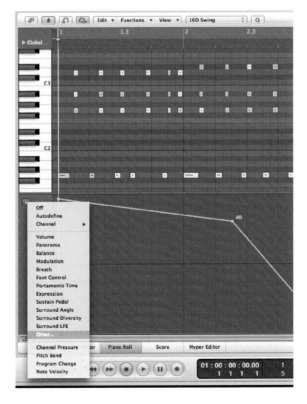

Use the Other option to display a MIDI controller type outside the main functions like volume and pan.

message types. For example, it might be that your software instrument or hardware synthesizer responds to MIDI controller message as a means of controlling filter cutoff, for example, or that it can be configured to work with certain controller message.

6.17 Hyper Draw in the arrange area

Although we're going to hop out of the Piano Roll editor for now, it's important to note the parity between Hyper Draw editing in the Piano Roll and the Hyper Draw facilities in the arrange area. In effect, the arrange area provides the full range of Hyper Draw features that we find in the Piano Roll editor all conveniently accessible via the View menu. Immediately, you should notice small previews of the moves we've made, but by selecting a region, choosing View > Hyper Draw, and selecting the controller type from the drop-down menu, we can see and edit the controller moves as if we were in the Piano Roll editor itself.

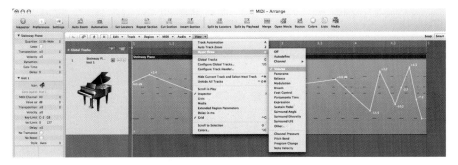

Using the View > Hyper Draw menu option, you can see and edit moves that you've created in the Piano Roll editor.

As you'll see, the Hyper Draw feature is very close to the functionality of Logic's Automation system – something that we'll take a closer look at in Chapter 8. In particular, what we're playing here is a form of region-based automation – in other words, data and "automation" moves that are directly stored within the MIDI data of a region.

6.18 Going further: the Piano Roll's edit and functions menus

One of the quickest ways of speeding up the editing process is to explore alternative methods of selecting and processing note events. For example, you might only be interested in notes after a given position in the region, notes that have been previously muted, or notes at the top or bottom of a group of chords. By applying an automated selection process, you can achieve complex

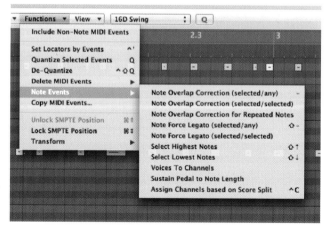

The edit and functions menu facilitates powerful tools both for the selection and processing of MIDI data, making MIDI editing both more efficient and more creative.

editing operations with relative speed and efficiency. Equally, various features in the functions menu allow you to select and manipulate MIDI information in ways you wouldn't immediately think of, but is, nevertheless, a gold mine of useful creative options and technical solutions to MIDI editing.

6.19 Intelligent selection: the edit menu

Besides the usual cut/copy and paste options, the edit menu contains a number of different selection criteria. To better understand the edit menu functions in the production process, though, let's look at some key musical applications of these features.

Select Equal Subpositions is a great tool for working with any rhythmic regions where the material spans several bars. On a hi-hat pattern, for example, we might want to accent particular beats across a number of consecutive bars. Start by selecting the appropriate notes in the first bar of the sequence, then go to the Edit > Select Equal Subpositions. Having activated the function you should now see the appropriate notes selected, which you can then increase the velocity of using the velocity tool.

Toggle Selection is a useful way of "inversely" selecting notes in the Piano Roll editor. Say, for example, you're interested in modifying a large number of notes (maybe the un-accented notes in the example of the hi-hat), but, at the same time, leaving a few notes unmodified. In this case, you're far better starting by selecting the notes that you don't want to modify, then toggling the selection (Shift + T) to invert the selection leaving the actual notes you want to work with selected.

Here's our simple hi-hat pattern; in the first bar we've selected the accent points that we'd like to pick up across all the bars in the region.

Using the Select Equal Subposition, we can add the corresponding notes in the subsequent bars quickly and easily. Now all we need to do is raise the velocity to complete the effect.

Toggle selection is a great function to work in conjunction with the note-muting feature. For example, say you want to split some MIDI information between two virtual instruments from the same "parent" region. Starting with the first region, use the mute tool to remove the notes that you aren't interested in hearing with this particular instrument. Now copy the region over to another instrument track, open it up, and select the muted notes using the Edit > Select Muted Notes. Un-mute the selected notes (M), toggle the selection (Shift + T), and then mute the newly selected notes. From one simple selection, you should have now made two parts with alternate sets of active notes.

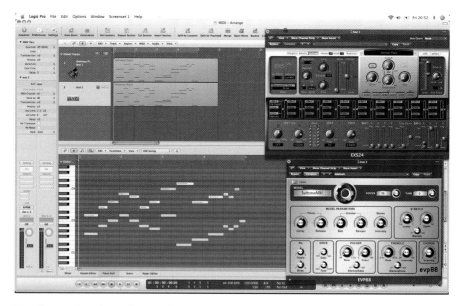

Toggling and muting selections between different instruments can be a useful way of distributing MIDI information across your arrangement.

Select Equal Events

Now and again it's possible for two MIDI events, of the same pitch, to end up on top of each other. This usually happens as a by-product of quantization, and, if you listen carefully, can be heard as a slightly "airiness" or "phasing" on the note in question. If you suspect this is happening, one quick way of rectifying this is to use the Select Equal Events option from the edit menu, and then hit delete on your keyboard. Although you might not see any immediate difference, Logic will have deleted these "double" events, with the note sounding a more natural way.

6.20 Functions: quick-and-easy note modifications

The functions menu includes a number of different tools both for selecting notes, and applying various transformations to MIDI data. The tools include some marvellous tools for dealing with music arrangement, especially with orchestral samples – either to create fluid legato passages, for example, or split played chords out into a series of parts.

Note Overlap Correction/Note Force Legato (Functions > Note Events)

Both these functions are interesting tools to work with when you're working with instrument samples – like violin or clarinet, for example – although they

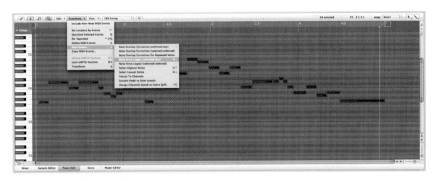

Using the Note Force Legato feature (Functions > Note Events > Note Force Legato), this MIDI region has had all its notes lengthened to meet the subsequent note.

can have a number of musical applications. Use Note Overlap Correction with a MIDI region being directed to a "monophonic" acoustic instrument, like an oboe or clarinet. In the real world, it is impossible for these instruments to play two notes at the same time, so any accidental note overruns (where one note has been held for too long) can affect the realism of the output. Note Overlap Correction, therefore, will remove any accidental overlapping of notes.

Alternatively, Note Force Legato is a great solution where you want the part to take on a fluid, expressive quality – like violins playing a lead line. With Note Force Legato, all the notes run into each other, irrespective of their original played length, producing a smoother overall line.

Select Highest Notes, Select Lowest Notes (Functions > Note Events)

By using the keyboard shortcut Shift + Up arrow or Shift + Down arrow, you can select the top and bottom line of a polyphonic region (in other words, a

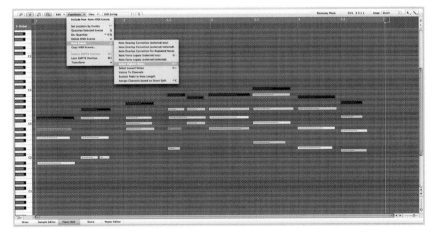

Use the Select Highest Notes or Select Lowest Notes to strip out the top or bottom line of a polyphonic MIDI performance. This could then be assigned to different instrument, for example.

167

region with multiple notes at any given point in time). One immediate application of this is dividing a played chord of the keyboard between three or more monophonic string instrument (violin, viola, and cello, for example). To do this simply, copy the three-part harmony over to the various instruments and use the Select Highest Notes, Select Lowest Notes accordingly, along with the mute function, to remove the required notes in each part.

Voices To Channels (Functions > Note Events)

Use the Voice To Channels feature as an intelligent, automatic means of splitting a polyphonic chord part into a series of unique voices. This end region will look the same, although on closer inspection you'll find that each voice has been assigned a different MIDI channel. To strip the data out, select the region and choose Region > Split/Demix > Demix by Event Channel. The polyphonic region should now be exploded into a number of different regions, one for each "voice" of the original part.

Knowledgebase 3

Using multi-output instruments

Both the EXS24 and Ultrabeat instruments are available in multi-output versions, in addition to their standard mono and stereo configurations. It's also possible that a number of other third-party software instruments in your collection – like FXpansion's BFD – might also come in multi-output versions. The notion of a multi-output is simply to provide additional outputs from the plug-in. Each output has its own fader, and, therefore, the possibility of adding additional effect plug-ins specific to the output in question, rather than the instrument as a whole. This is most often the case with drums, where you might wish to place different reverb, EQ, and compression settings on each different part of the kit.

To access the additional output, you'll need to ensure to instantiate the specific multi-output version of the plug-in. To add in additional aux channels, simply click on the + sign at the bottom of the instrument channel

Use the multi-output versions of instruments like the EXS24 and Ultrabeat so as to be able to process and mix individual elements from their sound output.

strip, with each subsequent press adding another output into the equation. To access the outputs, you'll need to route the appropriate signals using the plug-in's controls. On the EXS24, for example, the routing can be achieved via the EXS24 editor, while Ultrabeat's output options can be found next to each drum voice.

6.21 Step-time sequencing

Step-time sequencing dates back to the earliest analogue CV sequencers that wouldn't technically allow you to record "real-time" musical information (as Logic does), but instead store music as a series of 16–32 control voltages. Nowadays, however, step-time sequencing is still a valuable tool, mainly as a means of "stepping through" and creating a MIDI sequence piece-by-piece, rather than having to play a part by hand. If you're creating sequencer-driven music – with flowing arpeggiators and fast-moving basslines – then step-time sequencing arguably offers a more effective means of recording a part, without the need to quantize and meticulously edit your original sloppy performance.

To engage step-time recording in the Piano Roll editor, you'll need to activate the small MIDI IN icon in the top left-hand corner of the window. Whenever you press a MIDI note in step-time mode, Logic will place one or more (if you play a chord) notes at the current song position, and then advance one step forward at an amount defined by the current snap setting. Working in this way, you can quickly create a fast-moving sequencer line playing each note in sequence, but without having to play in time with the metronome.

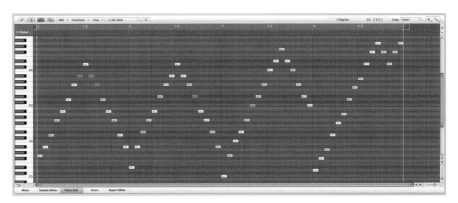

Here's an example of sequence created in step time, with a fast-moving 16th arpeggio effect.

As an alternative to using a MIDI keyboard, you can also make use of Logic's Caps Lock Keyboard (a feature that allows to play an instrument using your Qwerty keyboard, which can be called up by pressing Caps Lock down) as well as the dedicated Step Input Keyboard found under Options > Step Input

Keyboard. The Step Input Keyboard is a good solution for drawing note events using the mouse, with a clear horizontal representation of the musical keyboard, selectable note lengths, velocity settings, and so on.

Use the Step Input Keyboard and Caps Lock Keyboard as alternative methods of note entry for step-time recording.

6.22 The Hyper editor

Although we've spent some detailed time exploring Logic's main MIDI editor – the Piano Roll – it's worth familiarizing yourself with some of the other MIDI editors and what they can individually offer to the MIDI editing process. Having seen Hyper Draw, therefore, the specific Hyper editor shouldn't come as a great surprise. As with Hyper Draw, the main focus of the Hyper editor is in the editing of MIDI controller data – like volume, pan or modulation – although the specific benefit here is that we get to see a number of track lanes at the same time.

Editing and creating MIDI controller data in the Hyper editor

One quality that should become apparent about the Hyper editor is that it prefers to display its controller data as a series of bar graphs (usually on a 16th grid), rather than the line + node approach of Hyper Draw. Technically, this creates a slightly different approach to editing, although the end effect is much the same. It does, however, also make for an excellent source of pseudo-random 16th Sample-and-Hold movement, where a pattern appears to pulse up and down in time with the music. Try doing this with the modulation controller, for example, mapped to filter cutoff on the ES2 synthesizer.

For more generalized movements, you can use the pencil tool to draw in a series of stepped MIDI events – maybe creating a quick ramping up and down of volume, for example. Using the Line Tool you can also draw more gradual movements up and down, in the same way as the line feature in Hyper Draw.

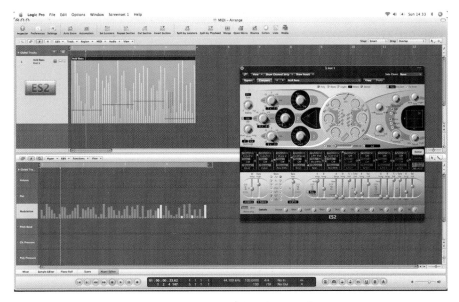

The Hyper editor is particularly good at stepped 16th controller movements.

Knowledgebase 4

Demixing

The Demix feature is a simple way of exploding a MIDI region into its constituent parts – this could be based on note pitch, for example, or by the Event Channel. To Demix a region first select it, then choose Region > Split/Demix > Demix by Event Channel or Region > Split/Demix > Demix by note pitch. The note pitch option, for example, is a good solution to use in combination with a MIDI drum loop, where a number of different lines (Hi-Hat, Toms, and so on) have been sequenced into the same region. By demixing the region, you'll create individual parts for each part of the kit, which you can then label Kick Drum, or Snare, for example, accordingly.

The Event Channel option works well combined with the Functions > Note Events > Voices to Channels feature in the Piano Roll, where a polyphonic part has been split into a number of different MIDI voices.

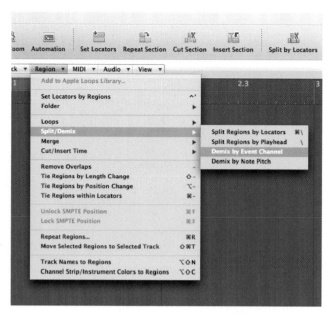

Demixing allows you to explode a single region into a number of separate parts based on note pitch or MIDI channels.

Adding new controller types and Hyper sets

To add a new controller type other than the ones in the pre-existing set, simply double-click in the blank space beneath the last controller assignment, effectively creating a new controller track lane. Opening up the inspector, you can then define parameters in relation to the controller lane, including the pen width, grid, and most importantly, the MIDI status type (program, controller, and so on), controller number (if applicable) and MIDI channel.

The particular arrangement and order of controller track lanes can be saved of as part of your own, custom-designed Hyper set. You could, for example, create a Hyper set that specifically relates to the MIDI control of a particular instrument plug-in, or a simplified set based on mixing. To create or delete a Hyper set, go to the Hyper menu and select the corresponding option, or alternatively use the dedicated Hyper set selector (where you can also switch sets) embedded into the inspector.

Editing drums using the Hyper editor

As well as editing controller data, the Hyper editor can also be used to edit and create note data, using the same series of 16th "blocks." As such, the Hyper editor is a good solution for drum programming, with the height of each block providing quick-and-easy velocity adjustment, while the pencil tool allows to draw in straight lines of 16th events. Usefully, Logic provides a pre-created GM Drum Kit Hyper Set, specifically designed for drum programming.

Use the GM Drum Kit Hyper Set as a useful way of creating and editing drum sequences. Note that you can also create your own Hyper sets to match the mapping of virtual instruments like BFD.

As most drum kits are mapped to the GM standard, this should be good for the majority of applications, although you can always program your own drum-based Hyper sets for plug-ins like BFD or Battery.

6.23 Score editor

The Score editor is the final MIDI editor selectable from the bottom row of tabs. Obviously, this editor is an appropriate choice if you have an ability to read printed music, although it has several key limitations for serious and detailed MIDI editing tasks. As a score is effectively "visually" quantized, the Score editor provides us with a relative coarse view of a note's placement and, while we're at it, its length. This makes any change beyond pitch manipulation somewhat flawed.

Of course, the real benefit in including the Score editor is in the preparation of music for musicians to play. Music preparation, as it's called, is a process we'll be taking a closer look at in Chapter 11.

6.24 Event List

The last MIDI editor that we're going to take a look at is the Event List editor, which displays the MIDI information in pure "data" form. The Event list can, at first, feel a little daunting and perplexing, and certainly not comparable to the visually driven approach to MIDI editing defined in the arrange area, Piano Roll, Score and Hyper editor. However, there are times when it's

The Score editor is a quick-and-easy way of editing pitch-based information, but provides us with little control or information on positional data.

important to get "down-and-dirty" with the data – understanding precisely where an event occurs for example (down to tick-level accuracy), or the exact parameter level with a controller.

The Event List editor can be opened as part of Logic general "list" displays – including markers, tempo and time signature lists – that can be opened up

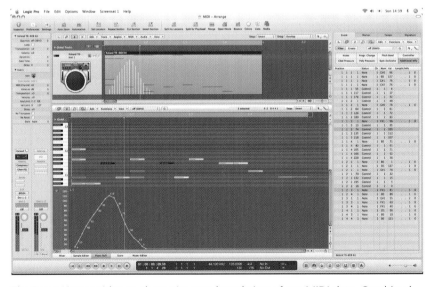

The Event List provides an alternative text-based view of our MIDI data. Combined with the other editors, the Event List can be an indispensable tool to accurately understand the content of your MIDI region.

on the right-hand side of the arrange window. This is an effective solution, as often, you'll want to combine the Event List with other MIDI editors like the Piano Roll or Hyper editor. To open the Lists display press E on your keyboard, or View > Lists, with the Event list accessible from the Event tab at the top of the display.

Filtering events

Given the range and quantity of MIDI data that can be included in a region, it is often necessary to filter different types of MIDI data out of the Event List. By clicking on the filter tab, you can then switch in and out various MIDI events, including notes, program change, controller, and so on. Using the filter can also be an interesting way of selectively deleting certain types of MIDI data. For example, you could filter note data so as to select and then delete extraneous MIDI data (controllers and so on) that you might not be interested in.

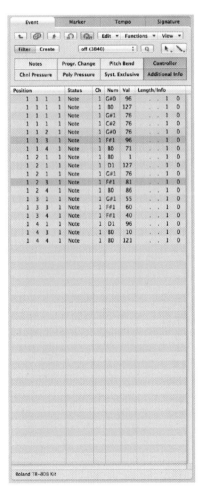

Use the filter tab as means of filtering out certain types of MIDI data.

Modifying and adding events

Editing data in the Event List is as simple as clicking on the appropriate number and either dragging up and down, or double clicking and entering a new value numerically. Numerical entry can be a surprisingly effective way of positioning notes and other MIDI events both quickly and accurately. You can also add new events, based on the current song position, by clicking on the Create tab and then selecting an accompanying message type. For program changes, or static controller levels, the Event List shines as a quick effective solution with the minimum of fuss.

Knowledgebase 5

EXS24 data management

Although it's quite possible to operate and work with Logic without an understanding of the EXS24's occasionally complex means of data management, it can become increasingly important as your own user-installed collection of sampler instruments begins to swell. The key to understanding where your data is, and how this is managed, is in differentiating the different sources of sample instruments – in other words, original library data installed with logic, user-installed and created content, and data that specifically relates to the current project. Although this can seem complicated at first, a little time investment will allow you to better access, organize, and archive this important data.

Factory content that was installed with Logic can be found under the path Library/Application Support/Logic/Sampler Instruments, with the associated sample files (remember an instrument file is separate to the sample data) under Library/Application Support/Logic/EXS Factory Samples. Your data will be saved in your user file, under *User Name*/Library/Application Support/Logic/Sampler Instruments. Whenever Logic is loaded, it will scan these two folders and build up an Instrument list based on the contents. This also means that you can actively use the Finder to organize the instrument files as you see fit, creating custom list of folders, for example, or even moving or duplicating favourite instruments into a single folder.

In addition to these two main folders, Logic will also take a look inside your Song folder to see if any EXS instruments are stored there. This may well be the case if you've inherited a project from another user who's backed-up a complete set of assets with their project archive. It should be pointed out, though, that these project-specific instruments can't be accessed in other projects. To do this, you'll need to copy the instrument files into the appropriate Library folder.

Samples for any of these instruments can reside on any drive, in any folder, with Logic scanning for the appropriate samples as it loads the instruments. The two immediate implications here, though, are both the scattershot sample organization that can occur, and the speed implication of a fragmented collection of samples. If possible, therefore, try to organize a large part of your sample data (you can leave the EXS Factory data) into a neatly separated folder, partition, or even better, a separate drive.

You can use the Finder to actively manage the organization and arrangement of your sampler instruments and how these appear in the EXS24 instrument menu.

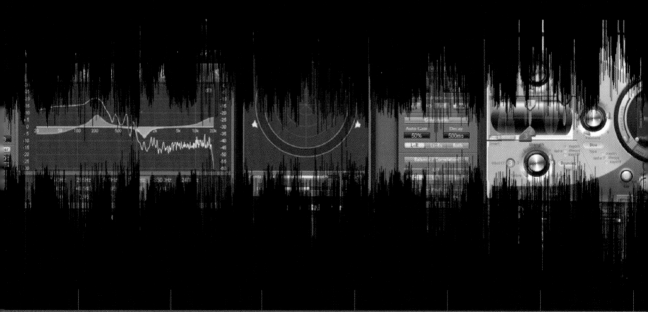

Creative sound design

7.1 Introduction

For anybody actively engaged in the more creative aspects of designing and modifying sound – say, a film sound designer, for example, or a musician working in the realms of electronica – there can be little doubt that Logic represents one of the most exciting applications to work in. Right out of the box, the modern-day sound designer has access to a huge range or synthesizers, samplers, and signal processors without having to go anywhere near any third-party products. Indeed, it's highly likely that your whole sound design project (whatever form it takes) could be realistically achieved entirely within the realms of Logic.

Having worked on the basics of using virtual instruments and MIDI in the last chapter, this chapter goes on to look at the creative applications of Logic's virtual technology to shape and craft any number of interesting sounds – whether you're using its versatile set of synthesizers (ES2, Sculpture, and so on), the pure sonic muscle of the EXS24, or the range of esoteric signal processors. Most importantly however, making your own sounds (rather than just simply relying on presets) can be a real opportunity to stamp your own identity onto a particular audio product, and ultimately, make your work stand out from the crowd!

7.2 Logic's synthesizers

Beyond some of the basic synthesizers we looked at in Chapter 6 – the ES E, ES P and the ES M synthesizers – Logic also includes a number of other synthesizer plug-ins, each dedicated to a different branch of synthesis and capable of an enormously wide palate of potential sounds. So, let's take a look at the three remaining synthesizer plug-ins, and what particular strengths they have to offer to the sound designer.

ES2

The ES2 is Logic's definitive answer to the good-old polyphonic analogue subtractive synth – like Roland's Jupiter 8, for example, or Sequential Circuits'

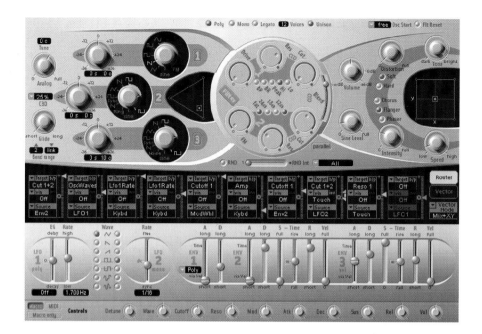

Prophet 5. However, this only forms part of the picture, as the ES2 has plenty of further tricks up its sleeve, namely, Vector Synthesis, Wavetable synthesis, and a host of other sound design tricks borrowed from many of the classic digital synths of our time: Waldorf's MicroWave, the Korg Wavestation and the Prophet VS. Make no mistake, this a tremendously versatile synth – not the easiest of plug-ins to get your head around, but one that will consistently surprise you with what it has to offer.

EFM1

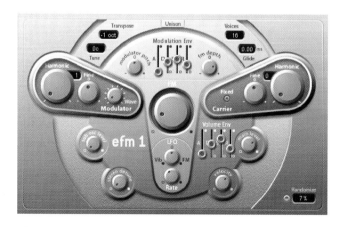

The EFM1, although technically part of Logic's more budget-orientated synths, is actually an interesting and valid example of FM synthesis in action.

For those of you that remember, FM synthesis was the primary system for synthesis used by Yamaha in the 1980s, and was at the heart of their infamous DX7. Although technology has moved on significantly since the golden days of FM, it can still be a valid source of sounds impossible to achieve elsewhere.

Sculpture

Sculpture is easily the most cutting-edge synthesizer in Logic's collection, and represents one of the most exciting developments in synthesis in the last few years – that of Component Modelling. Unlike synths like the ES2 and EFM1, Sculpture works on the basis a series of complex mathematical models describing the precise acoustic behaviour of a particular instrument. For example, rather than synthesizing the sound of a guitar by selecting a "similar" waveshape and approximating its behaviour with filters and envelopes, Sculpture precisely understands what happens as a string is plucked (using a variety of different plucking mechanisms) and what happens as the sound dies away. Not surprisingly therefore, Sculpture produces sounds impossible to achieve on other synthesizers.

7.3 Understanding the ES2

For most conventional music and sound design activities, the ES2 will probably be your main synthesizer of choice – both for the simplicity of working its interface, alongside the range of subtractive synthesizer sounds so eminently useable in contemporary electronic productions. The full interface itself is divided into the five key areas, illustrating the principal audio signal flow through the synthesizer (Oscillator > Filter > Amplifier, from left to right) alongside the interaction of modulation sources like LFOs (low-frequency oscillators) and ADSR (Attack, Decay, Sustain and Release) envelope generators. An additional "Macro only" mode – selectable from the bottom of the ES2 – presents a minimized control panel, which might be suitable if you're only interested in tweaking presets.

A The oscillators

Towards the top left-hand corner to the interface you'll find the controls for the three oscillators included within the ES2. The oscillators form the starting point of sounds created on the ES2 – the raw, harmonically rich waveshapes from which the filter gauges out harmonics, and big part of the identity of the sound. Thankfully, the ES2 includes a number of different principal waveshapes, and even a wavetable (more on this later) as well as a unique triangular vector mixer for the purpose of blending the oscillators together.

B The filters

Once the waveshapes have been selected and mixed, they are passed on to the circular filter section. As previously mentioned, the filters subtract harmonics from the original waveshapes – making the sound darker or thinner depending on the exact filter settings. On a musical level, the filters are important for creating the "juicy" sound of analogue synthesis, as best exemplified in sounds like the classic TB-303 bassline used in dance music. However, what's unique about the ES2's filter section is the fact that it contains two completely independent filters, opening up any number of different effects only possible with fully modular synthesizers.

Knowledgebase 1

Waveshapes

As you've seen with the oscillator section of the ES2, waveshapes can have a big effect on the type of sound you can produce. However, what defines the sound of each waveshape, and how best can they be used?

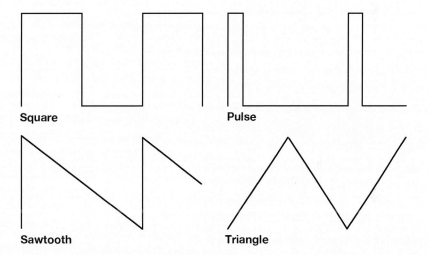

Of all the four principal waveshapes, the sawtooth produces the richest amount of harmonics, and is therefore the most popular choice for creating many of the classic synthesizers sounds. From warm pads, to gutsy synth basslines, the sawtooth wave can really excel, giving the sound plenty of presence, depth and bite! Be warned though, overuse of the sawtooth can make a track appear rather stoggy and one-dimensional, so it's worth exploring other variants and the sounds they can produce.

Compared to the sawtooth, a square wave produces a distinctive hollow and open sound to it, by virtue of the fact that it only contains odd harmonics from the harmonic series (a sawtooth contains both odd and even). Acoustically speaking, this makes the square wave similar to the tone of an instrument like the clarinet; although it's also useful for bass sounds where it can sound "big" without grabbing large parts of the sound spectrum (and eating up your mix, therefore) like the sawtooth.

The pulse wave, a variation of the square wave, is created by changing the duty-cycle of a square wave, effectively producing a wave that both sounds and looks thinner and more nasal than the square wave. For example, the ES2's Oscillators 2 and 3 both feature a fully variable square wave, allowing you to morph from a conventional hollow square wave to the thinnest of pulse waves. One common technique is to modulate the wave's width using an LFO, producing an effect otherwise known as pulse-width modulation (or PWM for short). Applied onto a pad sound, PWM can be a great way of adding a slash of warmth, somewhat reminiscent to the sound of chorusing.

A Triangle wave creates the purest sound of the four types, containing just enough harmonic material to make it still worth putting through a filter. Triangle waves can be a great choice for a sub-oscillator, tuned an octave down from the main oscillator. Try this on a sawtooth bass, creating a woofer-shaking subsonic to the patch.

C Amplifier and effects

The amplifier section – or in other words, the volume knob – controls the level of sound leaving the ES2 outputs. This section also includes a distortion unit (great for producing dirty, grungy, synth sounds) and a simple chorus/flanger/phaser unit for warming up any lifeless pad sounds (for example, this was a popular trick used on many of the older Roland synthesizers).

D Modulation matrix

Left to their own devices, the Oscillators and Filter won't produce much interest to the sound by virtue of the fact that they are static – in other words, the sound doesn't change with time. By adding in modulation, however, we can start to add interest, dynamics and movement into our patches. One simple example of modulation could be the process of linking keyboard velocity to the cutoff point of the filter – now, the harder you play, the brighter (or less filtered) the sound will appear. Other principal modulation sources include Envelopes, which shape the ADSR characteristics of a sound over time, and LFOs, which can be used to add effects like vibrato or wah-wah.

As you'd expect, the ES2 features one of the most comprehensive and flexible modulation matrices going, with access to almost every facet of the ES2's programming architecture.

E LFOs and Envelopes

The bottom section of the interface contains the controls for the three main modulation sources, including two LFOs and three Envelope generators. The key concept to grasp here is that each LFO and Envelope has a slightly different design, allowing each to excel in different tasks. LFO 2, for example, can be synched to the tempo of the song, while Envelope 1 provides quick and easy access to AD envelopes, which is particularly useful for percussive sounds.

7.4 Working with oscillators

With an understanding of the overriding synthesis architecture of the ES2 under our belts, let's take a closer look at creating sounds on the ES2.

The first step in synthesizing your own sound on the ES2 should be to consider how you intend to deploy the various oscillators – do you, for example, just use a single oscillator, or use a number of oscillators richly detuned to create a thicker texture? Deactivating the oscillators can be quickly achieved by clicking on the numerical legend to the side of each waveshape pot, this will also have a corresponding reduction in the amount of CPU processing used, an important consideration for those using G4 processors.

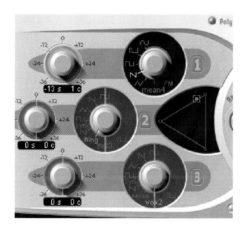

In this example, we've used just a single oscillator to create a simple synth bass sound.

As with many parts of the ES2, each oscillator can be seen as having some unique attributes, as well as features common to all three oscillators. Looking first at Oscillator 1, you should see a relatively standard set of synthesizer waveshapes: including Sawtooth, Pulse, Square, Triangle and Sine Waves. One of the more intriguing options includes the wavetable at the bottom of the dial (preset to a sine wave), and a fully variable FM option (for more information on this see Section 7.7, FM Synthesis, in Chapter 7). To audition the various waveshapes available as part of the wavetable, try clicking on the name and dragging the mouse up or down to scroll through the options.

Oscillators 2 and 3 add some further interest with the addition of a fully variable pulsewidth – morphing between a hollow square wave at the top of the dial, and an increasingly thin and reedy pulse as you move the pot clockwise. Another important addition is the provision of two synchronizable oscillators, using either a sawtooth or square waveshape, controlled from the "master" oscillator – oscillator 1. Going further still, oscillator 2 also includes the option to ring modulate oscillators 1 and 2, with oscillator 2 carrying the ring-modulated output. Oscillators 3's "special feature" is the provision of a noise waveshape – a great source for the random harmonic information so important to percussion sounds.

In situations where you're using more than one oscillator, you'll want to experiment with their respective tunings. One simple technique is to apply a slight detune, say ±4 cents, using the fine control on the left of the tuning dials. This should fatten up the sound you're producing – the equivalent of three violinists, for example, playing the same musical motif. The use of semitone shifts adds

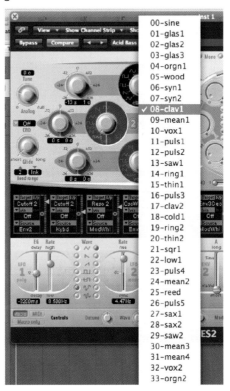

Oscillator 1 includes a full wavetable made up of a number of digital waveshapes. Try Ctrl-Clicking on the waveshape name to access the list as a menu.

Knowledgebase 2

Oscillator Sync

Let's get this clear from the start, Oscillator Sync has nothing to do with synchronizing the oscillators to the tempo of the track, although of course, you can achieve this with the ES2's LFO 2. Truly, oscillator sync is "synchronizing" the wavecycles of oscillator 2, with respect to the pitch oscillator 1 – in other words, the pitch of oscillator 2 is controlled by oscillator 1. At first this doesn't seem that exciting – simply controlling the pitch of one oscillator from another – however, by freeing oscillator 2's tuning controls from conventional tuning applications, we open the lid on a whole new world of exciting effects.

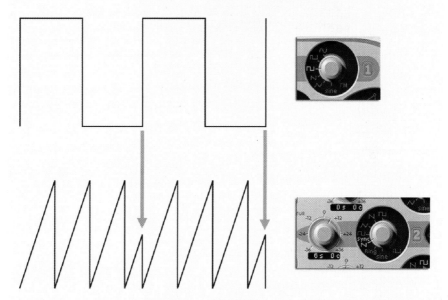

To fully understand oscillator sync, we clock what exactly happens to the pitch and timbre of oscillator 2 once it is placed in-sync with oscillator 1. The effect is most clearly audible when the initial pitch of oscillator 2 is set higher than oscillator 1, so try initially tuning the oscillator up 12 or more semitones. When the oscillator is placed in sync, you'll notice its pitch return to that of oscillator 1, although its timbre has become brighter and harsher – almost distorted, in fact. Looked at under the microscope of an oscilloscope (try rendering the synths output as audio file and taking a closer peek in the Sample Editor), you should see the waveshape being forcibly re-triggered, even though it begins its oscillation at a higher pitch. Therefore, it is this periodic re-triggering that we perceive as the final pitch of the oscillator.

Oscillator sync forms one of the most fascinating ways of distorting a waveshape, essentially creating a different version of the waveshape based on the relative pitch difference between the two oscillators. Although it can be used as a static effect, oscillator sync works best when the pitch of oscillator 2 is modulated in some way (try using Envelope 1 for a classic "sync sweep" lead sound), creating dramatic timbral sweeps even before you've reached the filter!

Oscillators 2 and 3 include a number of interesting waveshape options – from fully variable pulsewidth to ring modulation.

further interest and power to the patch. Try tuning an oscillator (set to triangle or square wave) down an octave (−12 semitones) from the main oscillators to create a sub-harmonic, or using third and fifth intervals (+3 or +4 for the third, +7 for the fifth) to create chord-like effects from a single note.

Beyond the basic detuning of the oscillators, there are a number of other features on the ES2 that can add width and warmth to the sound you're working with. The Analog control, for example, emulates the pseudo-random pitch and filter-cutoff drifts that occur within the imprecise workings of an analogue synthesizer. By increasing the Analog parameter therefore, you can add an increasing amount of pitch and cutoff frequency drift, effectively making the patch appear warmer and less sterile.

Activating the Unison mode, in either of the ES2's monophonic modes (mono or legato), takes the "fatness" of your ES2 patches to a whole new level. As with the Unison control on a traditional analogue synth, ES2's Unison stacks further virtual oscillators (on top of the three usually available in a patch) to create a thick, detuned, chorus-like effect. Going back to the string analogy, Unison is the equivalent of going from three violin players to a whole section, creating a sound that demands your attention in a mix. The actual width of the effect is defined in two parameters – the number of voices assigned to the unison, and the Analog parameter previously discussed. In Unison mode, therefore, Analog changes its role somewhat and simply sets the amount of detune between the "doubling" oscillators.

Unison, of course, is a staple feature of many of the expansive lead synth sounds found in trance music, although used in moderation, it can also be

Activating the Unison control will layer several "virtual" oscillators on top of the three you're already using in the patch. This creates a thicker and warmer output to the ES2.

great way of creating some super-phat basslines. But, like all good things, Unison comes at a cost – namely, your CPU resources!

7.5 Filters, amplifiers and modulation

With the initial pitch and timbre defined by our oscillators, then comes the turn of the filters and amplifiers to shape the raw sound into something that grows and develops over time. The Filter section – found in the large central dial of the ES2 – removes unwanted harmonics from the rich, waveform-only output from the oscillators. For example, starting with a "buzzy" collection of sawtooth waveshapes, you could apply a strong low-pass filter to remove the high-end harmonics, leaving you with a soft and warm pad sound. Alternatively, use any one of the other filter types and filter routing combinations – like band-pass, high-pass, or peak filter types – and you can achieve a real wealth of different timbres. Also, it's worth noting that the ES2 actually contains two independent filters – either allowing you to layer two different filter settings, or use two filters one after the other.

Knowledgebase 3

Filter Types

A large part of the "sound" we associate with a synthesizer is brought about by the performance of its filter. Filters come in many shapes and sizes, or more specifically, types and strengths, each with their own characteristics and applications.

The first thing to consider is the strength of the filter – which in the ES2's case is measured in three different decibel levels: 12, 24 and 36 dB. As you'd expect, the higher the decibel rating, the stronger the filter's ability to attenuate frequencies above or below the cutoff point. Interestingly though, a stronger filter isn't always the most desirable – most of the early Roland synths, for example, took a large part of their charm from the relatively "weak" 12 dB filters. A Moog, on the other hand, was famed for its steeper, 24 dB filter. On the whole, I tend to use the stronger filters on darker patches, whereas the classic, 12 dB filters work wonders on TB-303-inspired patches with lots of squelchy resonance.

The type of filter (Lo, Hi, Peak, BR, BP) defines the form of frequency attenuation taking place. The principal filter type is arguably a LPF, which attenuates frequencies above the frequency cutoff point – transforming the initially bright output of the oscillators into a more rounded overall tone. In most musical applications the LPF makes most sense, and there are strong similarities to what happens within the acoustic instruments. For example, a piano note will start initially bright (filter open) with high-frequency energy decaying as the note decays (filter gradually closes). An HPF, on the other hand, works in reverse so that frequencies below the cutoff point are attenuated. HPFs, therefore, are a great way of thinning sounds out, especially if the mix is starting to sound heavy and cluttered.

Band-pass and band-reject filters (labelled BP and BR on the ES2) are combination of LPFs and HPFs used to selectively modify a small band of frequencies. Band pass, for example, only lets a small band of frequencies pass through, producing a characteristically small sounding output. Band reject, on the other hand, leaves most of the audio bandwidth untouched, and instead attenuates just one "small" band of frequencies, much the same as a tightly Q'd dip on a parametric Equalizer. Conversely, a Peaking filter applies a sharp boost at the given cutoff point, again similar to the EQ comparison, but this time a boost rather than a cut.

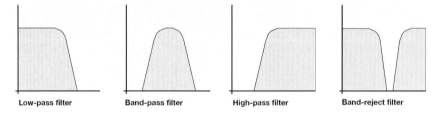

Low-pass filter   Band-pass filter   High-pass filter   Band-reject filter

Cutoff and resonance

The two main parameters with both the filters are Cutoff and Resonance. Cutoff, as the name suggests, is the point where the filter begins attenuating frequencies – so, in the case of a low-pass filter, a lowering of the filter will move the cutoff point slowly down the audio spectrum, producing an increasingly darker output. In effect, therefore, you're letting the low frequencies "pass," while the higher ones are attenuated.

Resonance, on the other hand, creates a small boost around the point of the frequency cutoff, arguably making the filter's attenuation more pronounced. The exact effect of resonance, though, can vary both between the filter types and the relative position of the cutoff. Using the example of a low-pass filter, a high-cutoff setting with some resonance applied will produce an increasing thin output (as the resonance boosts frequencies at the top end of the audio spectrum). Alternatively, resonance applied on a low-pass filter down the bottom of its scaling will produce an output with increasing amounts of low-end – an effect that can often be useful on bass sounds. Take the resonance all the way to its maximum setting, though, and the filter will self-oscillate (in other words, produce a high-pitched ringing sound almost like a sine wave) at a frequency defined by the filter cutoff position.

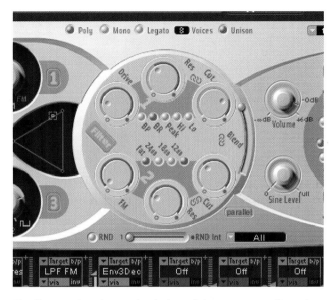

The filter section shapes the timbre of the raw output from the ES2's oscillators section. The two main parameters to interact with are filter cutoff and resonance.

Ultimately, the point to note with the filter is that it provides a large part of the "character" in a synth sound, probably best exemplified by the dripping, acidic resonance of a TB-303 filter in action on a bassline. All sounds, to a greater or lesser degree, will use the filter and if you can add some dynamic movement or expression into the patch (maybe responding to velocity, for example, use using an envelope and/or LFO), this character will become even more pronounced and interesting to the ear.

Changing volume over time

By default, the amplifier is controlled by Envelope 3, with the two remaining envelopes available for use with any other application you see fit. Envelopes allow you to manipulate the amplitude (or any other parameter, for that matter) over time – raising and lowering the given level for the duration for which the key is held down. The progression over time is defined by a number of key parameters: Attack, Decay, Sustain and Release (ADSR). Attack, like the majority of the other ADSR parameters, is a time measurement, so that a low setting produces a quick attack (useful for sounds that are more percussive in nature), while a high setting produces a graduated attack that, for example, might be applied to pads. Decay defines the time taken for the note to fall down to the sustain level, while release defines the final drop to silence after the key is released.

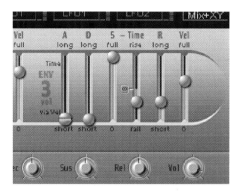

By default, Envelope 3 is patched to the amplifier, allowing you to shape the volume of the patch over time.

Looking at any sound in the natural world, it's easy to see how we can begin to define their qualities by the characteristics of an amplifier envelope. A piano, for example, would have a quick attack, slow decay, no sustain and a relatively quick release. On the other hand, a smooth evolving string sound might have a slow attack, slow release, medium-to-high sustain and a slow release. So, coupled with the basic waveform selection and the attenuation of the filter, we can really start to establish the defining properties and uniqueness of the patch we're creating.

Of course, alongside Envelope 3, the ES2 also contains two other envelopes, but to understand how these are used we need to look at the modulation matrix.

Bringing it all together: the modulation matrix

Tying all the elements together is the modulation matrix, which links a number of modulation sources in the ES2 (like the aforementioned Envelopes, LFOs, external MIDI controllers, and so on) to destinations within the synthesis architecture. The complexity and depth of the modulation matrix is what sets the ES2 apart even from many dedicated third-party synthesizers, and provides an almost unparalleled amount of flexibility and creative potential for synthesis. The only catch, however, is that certain elements of the ES2's control panel often need to be connected via the matrix before they have any effect on the sound!

Knowledgebase 4

Vector Synthesis

Vector synthesis, originally developed as part of the Prophet VS synthesizer, was first envisaged as a unique system for adding movement into a patch. In the case of the Prophet VS, it used a series of four oscillators, or sources, labelled A, B, C and D, which could be morphed between using a vector joystick control. By storing this movement as part of a vector envelope, the patch could then recall this dynamic movement creating timbral movement impossible to achieve elsewhere.

By switching the modulation matrix from Router to Vector you can activate the ES2's own vector synthesis features. The vector controls themselves work with both the triangular oscillator vector mixer (using three oscillators, rather than the Prophet VS's four), but can also make use of an additional vector controller in the form of the XY pad. By default, the XY pad has no assignment, but by using the X Target and Y Target parameters, you could, for example, assign the X and Y axis to cutoff and resonance – providing complete gestural control over two parameters on the one pad! To create a basic vector envelope, click on any of the vector nodes and position both the oscillator mixer and the assigned XY pad. Now select the other two nodes in turn, ensuring each has a different position for the oscillator mixer and XY pad.

To hear the vector envelope in action you'll need to uncheck the Solo Point switch – you should now hear the ES2 morphing between its various states. To add further nodes into the envelope, simply Shift-click on the time line. You can also set the time between any of the nodes by click-dragging on the various ms settings, making certain transitions quicker than others. Changing the loop mode will then allow the vector envelope to loop,

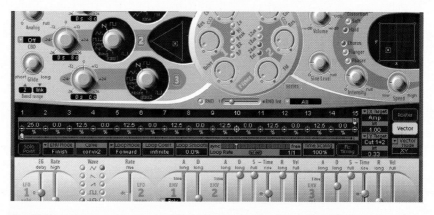

The ES2's vector envelope can add whole new levels of movement into your patch, by directly manipulating the oscillator mixer or XY pad.

based on the direction set and the relative position of the two markers – a sustain marker (where the envelope would hold, if no loop were set) back to the loop marker, indicating the full duration of the loop.

By exploring some of its deeper features, you can also turn the vector envelope into a surprisingly effective source of rhythmic pads and textures. Try changing the Loop Rate from "as set" to a tempo division like 1/2. Now the vector envelope appears to pulse in time with the track, a feature that could really be exploited with further nodes (up to 15 are allowed) and some interesting tuning and waveshape assignments of the respective oscillators. If you're in any doubt as to the effectiveness of this, try listening to some of the Vector Rhythm (Curve) presets as part of ES2 Legacy folder.

Looking more closely at the matrix, we can see a number of different routing paths (10 in total) indicated by the line of blue boxes. Each routing connection will have a target parameter (which could be a parameter like Filter Cutoff or Pan), a modulation source, and the amount of modulation that can be applied. Just a quick glance through the respective menus for both the target and source reveals the depth of the ES2, with over 40 program targets and over 20 possible modulation sources, with everything from standard modulation sources, like LFOs and Envelopes, to random generators (RndNO1 and RndNO2) and external MIDI controllers.

The ES2 modulation matrix allows us to map up to 10 modulation sources (LFO, Envelopes, and so on) to different destinations (Filter cutoff, and so on) within the synthesizer's architecture.

Some of the principal connections you might want to explore involve controlling the filter from modulators like Envelopes 1 and 2, and well as the two LFOs. The LFO provides a "periodic" form of modulation – in other words, a repeating continual shape or pattern that modulates the parameter, rather than an ADSR-type modulation with fixed stages. Arguably the best example of this is vibrato – where an LFO would be routed to the Pitch123 target. However, LFOs can also be an excellent source of Tremolo-type effects (routed to Amp) or a wobbling wah-wah (routed to Cut 1 + 2).

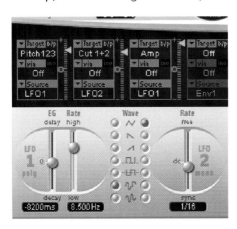

Each of the two LFOs can be set to a variety of different modulation shapes, each creating a unique form of modulation effect. For example, the sawtooth waveshape, or the stepped sample-and-hold waveshape can be great for sequencer-like effects, especially attached the filter. In that respect, it's well worth noting that, in its below "DC" setting, LFO 2 can also be set to sync with the project's tempo.

7.6 Global parameters and output effects

As part of the ES2 global parameters to the right of the filter dial we find a number of useful tools to add further layers of interest to the sound. The Sine

Level, for example, adds an additional sine wave element to sound based on the pitch of oscillator 1. This is a useful tool for general "bass fattening", especially if the output of your filter has left your patch sounding slightly weak. If you want more grit, though, the inbuilt distortion unit, which is switchable from hard to soft operation, helps give aggressive synth sounds considerably more edge. Even in relatively soft settings, with small amounts bled in, the distortion can make a patch sound slightly more analogue than its digital origins would suggest!

The Modulation Effects section provides inbuilt chorus, flanging, and phasing effects, with a variable intensity and speed. All of these are useful warming effects, adding an extra sense of depth, stereo width, and movement to your ES2 patch that really seems to suit pad-like sounds.

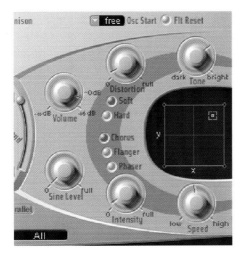

The output effects are a useful means of adding extra warmth (using the chorus unit) or grunge (using the distortion) to the ES2 patches.

7.7 EFM1 and frequency modulation (FM) synthesis

Subtractive synthesis and the features of the ES2 might cover a large part of the sounds used in contemporary music production, but there are other methods of synthesis – like frequency modulation, or FM – with plenty to offer. EFM1 is Logic's solution for FM synthesis (although FM also makes some honourable appearance in the ES2) using the same principles carved out in Yamaha's infamous DX range of synthesizers back in the 1980s. Although the EFM1 lacks some of the depth and complexity of a full-blown seven operator DX7, or Native Instruments' FM7, it remains a valuable insight into a unique form of sound creation and a valuable contrast to the sounds produced by the ES2 and Logic's other subtractive synthesizers.

Knowledgebase 5

ES2 Filter Routing

One of the most intriguing aspects of the ES2 is its immensely flexible filter routing system. Essentially, the two filters can be connected either in series – so that sound flows sequentially from one to the other – or in parallel, where the sound is split and fed to each filter individually. To switch between the two filter modes, simply click on the filter's Parallel or Series switch accordingly, and enjoy the sight of the revolving filter controls! Taking this up another level, we then have the Blend parameter, alongside the Drive control, both adding a whole new level of interaction based on the current filter arrangement and its relative position.

Looking first at the Parallel routing (where sound is split and sent through to two filters individually) the blend parameter can be seen as a simple crossfade parameter, making one filter's output more prominent that the other in the final mix. In serial mode the concept is a little trickiest to understand, although not impossible. In its central position, Blend creates a true serial configuration with oscillator 1's output being passed to the oscillator 2's. Either side of the line, however, an amount of either Filter 1 or 2 is progressively bypassed, effectively creating a bias towards one of the two filters. Of course, at either extremes of the Blend parameter, the output is exclusively Filter 1 or 2 respectively.

The Drive parameter adds polyphonic distortion either before or during the filter's operation (again, based on the filter's routing). Polyphonic drive differs from the conventional distortion (available in the amplifier section) in that it distorts each voice individually – in effect, using a different distortion

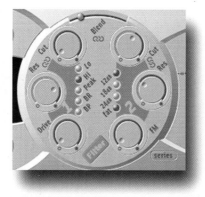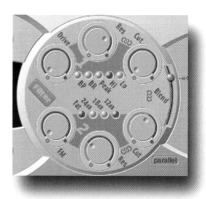

Here are two ES2 synthesizers, each set to one of the two different filter routings – parallel or serial. Using the blend and drive parameters adds further possibilities to how you can use and abuse the ES2's filters.

module for each note played! In short, this means that a chord can be given a little "grit", rather than collapsing in intermodulate distortion, as it would in a conventional single distortion module setting. In a Parallel routings, the drive is placed before the two filters, effectively giving them slightly more harmonic material to play with. In Serial configurations, the Drive in generally placed after Filter 1 (although the exact interaction with filter 2 varies given the blend parameter) making the distortion effect more pronounced, especially on high resonance settings.

Not surprisingly, the heart of FM synthesis, and the sounds produced by the EFM1, centre on the process of FM – but what is it, and how can it be used to produce musically useful results? The process of FM requires two oscillators – one oscillator acting as a Carrier, and the other oscillator acting as a Modulator. You can see this clearly on the EFM1 front panel, with the carrier on the right-hand side of the panel, and the modulator to the left-hand side. The relationship between the two is that the modulator is modulating the pitch, or frequency, of the carrier, with the carrier oscillator being the only oscillator present in the audio output.

In many ways, the simple modulator/carrier configuration we see as a basis of the EFM1 bears a fundamental similarity to an LFO modulating the frequency of an oscillator on the ES2. As the LFO is working below the audio spectrum (around about 6 Hz), we hear the effect of this modulation as vibrato, with stronger amounts of modulation creating an increased amount of pitch wobble.

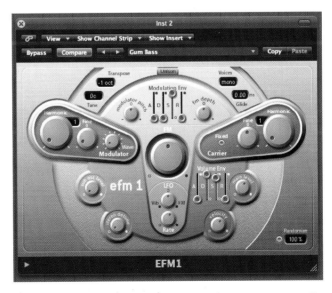

By increasing the amount of FM we create a more harmonically rich output, in much the same way as opening a filter on a conventional subtractive synth.

However, in the case the EFM, the modulator is working right up in the audio spectrum, producing an effect quite unlike vibrato. By modulating pitch using an oscillator in the audio spectrum, we effectively distort the original wave-shape, creating new harmonic components – the stronger the amount of modulation (set with the large central FM parameter) the greater the amount of the distortion and hence the richer the eventual harmonic output.

One of the most important relationships to understand in FM synthesis is the ratio between the pitch of the carrier and the modulator. On the EFM1, the tuning of the modulator and carrier is established using the Harmonic control on either side of the interface. Try setting the carrier to the first harmonic and experiment with the modulator on first, second and third harmonics. With both carrier and modulator set on the first harmonic, the sound will approximate a sawtooth, with a collection of both odd and even harmonics. Moving the modulator up to the second harmonic will produce a more hollow and open timbre, with an output largely made up of odd harmonics. Going further still, this time up to the third harmonic, the sound becomes more bell-like, by virtue of the fact that you're using an odd harmonic relationship.

Changing the relative tuning of the modulator and carrier changes the resultant harmonics produced as the FM knob is increased.

Modulation on the EFM1

Okay, we've established the basics of a "static" FM sound – namely the tuning ratios between the carrier and modulator, and FM depth – but how do we go about adding movement or interest into the patch? Let's first consider trying to create a basic filter sweep. On an analogue synthesizer this would be achieved using an envelope patched to modulate filter cutoff, with the envelope opening or closing the filter over time. In the EFM1's case, we can use the Modulation Envelope towards the top of the interface in conjunction with the

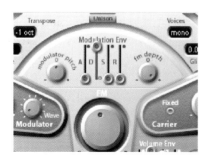

Use the Modulation Envelope and FM Depth parameter to modulate the large FM dial over time. The effect of this modulation would approximate a filter sweep on a traditional analogue synthesizer.

FM Depth parameter to the right of it. By setting up an appropriate envelope (quick attack, graduated decay, no sustain) with FM depth set to its maximum setting and the large FM dial set to its centre position, the EFM1 should then gradually reduce the amount of FM as the key is held down.

Knowledgebase 6

Wavetable Synthesis

Wolfgang Palm originally developed Wavetable synthesis in the early 1980s as part of his groundbreaking work on the PPG Wave synthesizer. The principle was that an "audio event" – like a filter opening – could be broken down and stored in a Wavetable constructed from 64 single-cycle waves. By stepping through these wave cycles, and interpolating between the various settings, the PPG could create unique and interesting textures, often characterized by a glassy, shimmering tone. Although PPG eventually went bust in the late 1980s, Palm's intriguing Wavetable system continued to live on (at least for a few more years!) courtesy of Waldorf's Microwave and Microwave II synthesizers.

By including its own digital Wavetable, the ES2 pays homage to this late great synthesizer, and, thanks to its flexible modulation matrix, can recreate many of the sounds that the PPG wave was so famous for. If you want to experiment with Wavetable synthesis, the trick is to select OscWaves as a modulation target, and ensure the oscillators are each set to their Wavetable

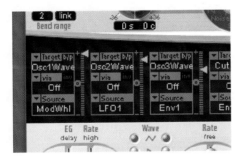

By modulating the Wavetables within the ES2 you can create some interesting "moving" synth sounds reminiscent of the PPG Wave 2.3.

position. Choosing the exact type of modulation source (albeit an LFO, Envelope, the Modulation Wheel, or something like keyboard velocity) will then dictate the type of movement and the sound produced. Try using a slow LFO (0.10 Hz, for example) and gently modulate the Wavetable potion to create a shimmering digital pad. Alternatively, you can create an interesting glitch-like effect by modulating the Wavetable using a fast AD envelope.

To add further interest into the equation, you can also modulate each of the Wavetables individually, possibly using a different modulation source (LFO 1, LFO 2 and modwheel) on each. To do this you'll need to use OSC1Wave, OSC2Wave and OSC3Wave destinations, respectively.

Another form of modulation is the LFO, which, in the case of the EFM1, can be moved between being applied to pitch – to create vibrato – or routed to the FM depth. As we've seen in the last example, routing a modulator to FM depth produces an effect comparable with filter modulation. In this case, therefore, routing the LFO to FM creates a form of wah-wah effect, with the timbre of the EFM1's output wobbling over time. An entirely more chaotic form or modulation, though, is possible by routing the modulation envelope to the modulator pitch (using the dedicated pot). The effect is an extreme, and possibly dissonant, change in timbre, which can be greatly used in aggressive synth sweeps.

Refining your EFM1 patches

To further refine your patch, the EFM1 includes a number of different parameters, which don't differ too greatly from the features found on the ES2. The Volume Envelope for example, allows you to shape the amplitude over time – for example, try creating bell sounds using a quick attack, no sustain and a slow decay and release. The sub-oscillator, as on the ES2, doubles the synthesizer's output with a sub-oscillator an octave below, giving bass sounds a real punch. Finally, the Stereo Detune option effectively creates a complete double of your patch, slightly detuned and with a distribution (of the two engines) across the stereo image. Used subtly, this can be a simple solution for adding warmth, depth and width to any of your EFM1 patches.

7.8 Component modelling: Sculpture

When it comes to cutting-edge sound design, Sculpture represents one of the most exciting and perplexing parts of the Logic, and also a real indicator of a future development of synthesis, and a component that may well become increasingly dominant in sound production as a whole. Sculpture's greatest strength – but equally the reason why it takes so much time to master – is that it throws out the conventional "rule book" of synthesis. You won't find traditional features like oscillators or amplifiers (although filters do get a minor input!), but instead parameters like Media Loss and Exciter Objects, producing sounds that are almost impossible to achieve elsewhere!

Like an increasing number of virtual instrument – including Arturia's Brass and Applied Acoustic Systems' String Studio – Sculpture is built on the concept of physical modelling, or, more specifically, component modelling. The basis of all these instruments is a complex mathematical model of an instrument, with a precise understanding of how sonic variables (the materials the instrument is constructed from, and so on) and various interactions (in other words, how it is played) affect the sound the instrument produces. In the case of Arturia's brass, for example, the "model" is based on brass instruments like trumpet and trombone, while in the example of Sculpture we have the multitude of possibly musical outcomes that result from interactions with a string!

7.9 Objects

The starting point of patch in Sculpture is to consider how your instrument is constructed, and the means in which it will be played. Indeed, in the first part of designing your sound, you will spend a lot of time moving between two key areas of Sculpture's interface: the Material Pad, in the centre on the control panel, and three different Exciter Objects, all interacting with each other and having some fundamental impact on the type of sound being produced.

Let's first look at the concept of Objects. To produce any note from a real instrument we need to introduce energy into it – for example, blowing across the reed of a clarinet or using a stick to hit a snare drum. Correspondingly, in Sculpture we have a choice of up to three Objects that can be introduced to the string to begin creating a note: it could be hit, pluck, struck, or blown (to name but a few of our potential objects), with each different object producing

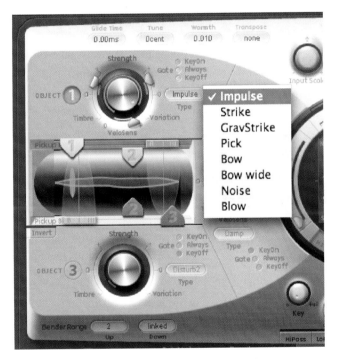

Sculpture's string can be played in a number of different ways, defined by the Type parameter of each object. For example, Object 1 includes Picking, Bowing and a number of other "interactions."

remarkably different results. What this immediately means is that you need to have at least one object active to create any audio output whatsoever. Also, by having more than one object, you can create some weird-and-wonderful effects like a plucked string being gently dampened half way up its length, or complete audio chaos with a number of objects all playing their part at the same time! As you can see, the best results will come with a considered application of these settings, rather than just throwing it all "into the pot."

Accompanying each exciter object you'll find another three key parameters – Strength, Timbre and Variation – that set some of the initial effects of the of the object against the string. Of course, with each different object model the precise impact of the three parameters varies, so it is well worth spending some time with each to understand the range of sounds and effects you can start to produce from them. If you've got the processing power, it might also be worth turning on the string animation by Ctrl-clicking on the thin green line towards the middle of the object controls.

Although you can produce some fantastic sounds with just the one object, the true potential and interest with Sculpture comes as you to add further objects into the equation. In particular, you'll notice that Objects 2 and 3 contain some of the more esoteric object options, like Disturb 2-Sided, which approximates the effect of a ring placed around the string, or Mass, which places a large weight at some point along the string. Note also that you can set the object to be triggered in one of three ways – KeyOn, Always and KeyOff – so that an object is only triggered after the key is released.

If you do use more than one object, it's worth paying close attention to the relative strength of the various objects. For example, using all three objects

Objects 2 and 3 contain a number of different string interactions, which can be used to enhance the basic settings established with Object 1 – like the string being picked and damped at the same time.

at maximum strength tends to create an overwhelming sound with little or no character. However, picking one key object (usually using Object 1) and augmenting this with subtle additions of other objects seems to produce the more interesting hybrid effects. For example, you could choose to use a Strike model on Object 1, which is then lightly Disturbed (the effect of placing an object on the string) using Object 2.

As important as the types of object used to initiate vibrations in the string, we also need to consider where along the string it is played. Take, for example, a nylon string guitar being played using a small pick. Played partway down the length of the string, just above the sound hole, results in a pleasing, rounded tone. However, if we were to play right up against the bridge, the tone will change dramatically, becoming harsher, with a thinner overall timbre. Just as with real life, therefore, we can position each of the three objects along our virtual string using the Object Positioning display in the centre of the Objects section.

Again, the best way of understanding how the models work, and their precise effect at each point along the string, is to experiment, although the general principle follows that a fuller sound emanates from the centre of string, while the extremities tend to produce a more nasal timbre. Adding in additional objects makes the variables ever more complex, although again, the general rule-of-thumb is that objects arranged closer together tend have a more pronounced effect on one another.

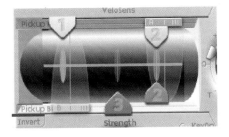

The position of an object can have a big effect on the string's vibrations. This interaction becomes even more complex with a greater amount of objects.

As part of the same object positioning display, you'll also find controls for each of Sculpture's virtual pickups. These work in much the same way as the electromagnetic pickups on an electric guitar, which of course, tend to output a different sound based on the relative position along the guitar string. In Sculpture's case, we're presented with two pickups that can be slid along the virtual string, contributing to the output in a number of different ways. To the far right of Sculpture's interface you'll also find the stereo control, which can define a stereo spread both for Sculpture's key range (moving from left to tight as you move up the keyboard) and between the two pickups. Try using a wide stereo with the two pickups so as to create pleasing width to Sculpture's output – an effect especially useful on pad sounds.

As with an electric guitar, the restive position of the pickups (A and B) will also influence the sound produced. Use the Spread control on the right-hand side of the interface to turn this into a stereo effect.

7.10 The string

Having spent so much time analysing the objects used to excite Sculpture's string, we've neglected to look at an equally fundamental part of creating sound in Sculpture – the design of the string itself. Again, relating principles back into real life, we can see that the way a string is made will have a big effect on the sound produced, as evident in the apparent difference between a nylon strung and steel strung acoustic guitar. By adjusting properties in the central Material Pad section, therefore, we can fine-tune the properties of our virtual string – morphing the sound between glassy, bell-like textures to warm, woody marimba tones, and plenty more besides!

The central X/Y controller of the Material pad sets the relative Inner Loss and Stiffness of the virtual string, although the precise qualities of this is better expressed by the small nylon, wood, steel and glass legends in the four corners of the XY pad. Try moving the small control ball and listening to the harmonic spectra as you move around the pad – for example, listen out for the brighter tones towards the bottom of the XY pad against the move towards more bell-like harmonics as you move from left to right. As with objects placed along the string, this simple pad has one of the biggest impacts on the sounds you produce from Sculpture, so it's well worth getting to know the various initial sounds possible from it.

Around the Material pad we can also find a number of other key parameters in relation to the design of the string: Resolution, Media Loss, and Tension Mod. The most important of these is the Media Loss parameter, which sets the dampening applied to the string based on its surrounding conditions. A more meaningful way of understanding this would be the "release" of the sound, so that with a low media loss setting, the sound will appear to decay over a long period of time, while quicker media loss settings tend to produce a shorter, more clipped sound.

Tension Mod corresponds to momentary fluctuations in pitch as the string is excited – the more the Tension Mod is applied, the greater these pitch fluctuations appear in the final output. Used subtly, Tension Mod can add a discrete sense of the natural pitch shifts that occur within a real acoustic

Using the central X/Y control will allow you to change the material properties of Sculpture's virtual string.

instrument, although care should always be taken not to overuse features like Tension Mod as it could make the instrument sound like it has poor intonation! Resolution, on the other hand, corresponds to the amount of harmonics used to generate Sculpture's output. Try experimenting with lower resolution settings, as this can be an interesting way of creating inharmonic overtones (great for bell sounds) and as a means of reducing Sculpture's CPU load.

Parameters around the Material pad set a number of other physical considerations. Media Loss, for example, could be thought of as a form of decay or release setting, governing the loss of energy over time.

Taking the string design up another level, we can also add variations to its design based on the relative key position, otherwise known as Keyscaling, and in its release phase. So, for example, as you play up the keyboard the sound could change from a warm, woody timbre to a bright, glassy string. Equally, Keyscale Media Loss could approximate the shortening of strings duration as you move up a piano keyboard, for example. To modify any of these settings, simply click on the Keyscale and Release bottom at the bottom of the Material Pad, and adjust the parameters (using the coloured indicators) as appropriate.

Using the Keyscale and Release control you can vary the properties of the string both in the release phase of the note, or in respect to keyboard position.

As a closing thought to the Object and String controls, try to remember that experimenting in component modelling synthesis is rather like re-learning the art of synthesis. So, rather than using tried and tested means of solving problems – like adjusting an envelope to change the dynamics of a sound, or a filter to change the timbre – you'll need to think "outside the box" and discover a new palette of solutions available in a component-modelling universe. Changing the amplitude envelope, for example, could involve a change or even an addition of an object onto the strings – a short decay, for example, made by physically dampening the string on the key off. Alternatively, adapting the string's Inner Loss properties, or simply less strength on the exciter, could make for a darker sound. Ultimately, it's all about new solutions to old problems.

7.11 Waveshapers and beyond

Theoretically, in a true component modelling synthesizer, the object and string parameters should be all that is required to create your eventual output. Indeed in Sculpture's case, it is highly likely that you'll spend the majority of your time working in these sections defining your sound, with the finished patch simply requiring a touch of EQ, compression, and reverb to set it in the mix. Thoughtfully though, Sculpture also provides a number of additional features, including Waveshapers, Filters, ADSR envelopes, Delay lines and EQs, to further enhance and modify the effects you can produce within the plug-in. Although these aren't necessarily integral to the concept of component modelling, they are highly useful additions.

Following Sculpture's internal signal path, we start with the Amplitude Envelope. The important thing to note here is that the amplitude envelope is post the component-modelling engine – its settings have no direct correlation to the sound produced, and merely act as a means of modifying what's already there. So, for example, extending the release phase has no effect if the string has stopped vibrating (instead, consider raising the Media Loss), but a slow attack could be a suitable solution for making a bowed string have an even more graduated entry.

The amplifier envelope allows you to change the volume characteristic over time, but remember, this is initially defined by the string and material properties.

The Waveshaper, which follows after the Amplitude Envelope, is one of the most useful additions to Sculpture's synthesis engine, essentially applying a polyphonic distortion effect (much the same as the ES2's filter distortion effect). In addition to applying a suitable valve-warming effect (try the Soft Saturation algorithm), the waveshaper is essential in producing synth-like tones from Sculpture, making the output of the component modelling engine much closer to the buzzy, harmonically rich output of a subtractive synth's oscillator. Of course, because the distortion is polyphonic (with each voice being distorted on an individual basis) even strong settings don't result in the sound collapsing in piles of un-musical inter-modulate distortion.

The Waveshaper adds grit or warmth to your Sculpture patch.

Working in close conjunction with the Waveshaper is the filter, which includes five different filter modes – Hipass, Lowpass, Peak, Bandpass and Notch. Given the added harmonic material created with the waveshaper, you can now use the filter to subtract harmonics, producing synth bass sounds, and so on. Also the filter can be a useful additional way of tapering the output from the component-modelling engine – maybe using a LPF to create a dark bell patch, or the high-pass filter (HPF) to take the low end out of a steel-string guitar patch. Remember, as with a conventional subtractive synth, the filter is also an ideal destination to control using the modulation wheel, an LFO, or an envelope, as we'll see in the Modulation and Morphing section.

Use the filter as an additional form of timbre control, with various different modes to change the relative effect achieved.

Three final modules provide the last refinements: a relatively simple Stereo Delay line effect (which isn't entirely dissimilar to Logic's Stereo Delay plug-in), a Body Equalizer and a Limiter. The body equalizer's simplest application is Low Mid Hi Model, used to cut or boost frequencies at 80 Hz, 2.5 kHz and 12 kHz, respectively. More advanced treatments are possible using one of the supplied instrument models, which are themselves drawn from impulse response recordings of the respective instrument bodies. In effect, what you have access to here is a detailed EQ profile of the characteristics of the instrument in question – it won't, for example, turn a harp sound into a cello simply through the application of the Cello's Impulse Response, but it does at least impart some of the instrument's tonal character.

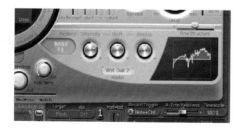

Use the Body EQ and the instrument-based IR to impose the timbral characteristics of an instrument onto your Sculpture patch.

Once loaded, the impulse response can be tweaked in a number of ways to best match the patch you're trying to create. Lowering the Intensity, for example, can be a useful way of softening the effect of the impulse response, imparting a subtle modification of tone rather than completely transforming it. Also, try experimenting with the Formant Shift and Formant Stretch parameters, allowing you to more carefully tune in the respective "lumps and bumps" of the impulse response profile with the tonal characteristics of what you're trying to create.

7.12 Modulation and morphing

As we saw in the ES2, modulation is essential for bringing a patch to life: making it more expressive in response to actions on keyboard, adding in LFOs to add effects like vibrato and tremolo, and using envelopes to shape the sound over time. As you'd expect, Sculpture is no exception to this rule, with a modulation section (towards the bottom of the interface) that rivals, if not exceeds, what is possible on the ES2. More importantly, the application of modulation on Sculpture has the potential to directly influence the specific properties of the instrument (string construction, object position, and so on) rather than using crude "signal modifiers" (filters, amplifiers, and so on) to approximate the same effect.

The bottom dark grey pane in Sculpture contains all the various controls for changing the way the instrument responds to a performance and behaves over time: including two LFOs, a Jitter generator, Vibrato, Velocity modulations, Controller routing, two Envelopes, a Morph pad, and a Morph pad vector envelope – enough to keep anybody happy! Pressing the various tabs allows you to step through the relevant sections of the basic modulators, while the Morph pad and envelopes hide some intriguing "contextual" menus and functions, like recordable envelopes and patch randomization.

The modulation section of the Sculpture adds an additional level of dynamics and movement to your patch.

Simple modulation can be accessed via the Vibrato and LFO 1 and 2 tabs. Vibrato permanently connects a spare LFO to the pitch of Sculpture, while LFO 1 and 2 can be wired to any number of destinations within the Sculpture patch – from the usually Filter Cutoff, and so on, to some of the more unusual component modelling parameters like Object Position or Pickup positioning. Exploring modulation with these parameters, in particular, can lead to some of the most musically useful and rewarding applications of Sculpture, with complex or subtle modifications in timbre that would be impossible to achieve on a traditional subtractive synthesizer. For example, try gently modulating the pickup position in a wide stereo pickup spread – the effect is a great way of adding subtle movement to a patch and palpable sense of "stereoness."

Routing the LFOs to the pickup position can be a great way of creating stereo movement in a patch.

For direct response from a MIDI performance, Sculpture features two controller paths – labelled Ctrl A and Ctrl B – which are usually assigned to the Mod Wheel (1) and Foot Pedal (4), respectively, but can be changed to other controllers as appropriate. Again, think about creative and expressive uses of these routings – one favourite technique is to use two spare MIDI controller faders to fully modulate the positioning of two of objects on the string. This can be a responsive way of playing Sculpture, almost like a guitarist subtly moving a pick to various point around the sound hole and combining this with selective damping or muting on the strings.

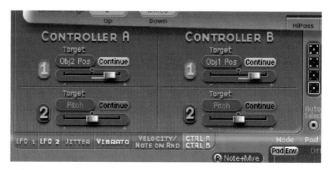

Use the external MIDI controller to manipulate the relative positioning of the objects on the string. This can add a surprising level of expression to your patch.

The Jitter, Velocity and Note On Random modulators can also be used to add a variable amounts of chaos, unpredictability and expression to your sound. Jitter, for example, can create small pseudo-random imperfections in the patch – try adding some Jitter routed to pitch, to recreate the slight pitch variations exhibited on some acoustic instruments. Velocity and Note On Random can also be effectively mapped to parameters in the Material pad, so that different velocities, or alternate notes, produce subtle (or extreme) variations in the properties of the instrument. In theory, you could stack all of these modulation sources up, so that Sculpture responds and modulates in a multifaceted and complex way.

Use Jitter to add pseudo-random modifications and drifts within your sound.

Although the modulation paths that we've discussed will start to bring some movement into your patch, probably the best modulation features of Sculpture are to be found in the envelopes and Morph pad. The envelopes can be used in a conventional way (mapping to Filter Cutoff, for example), or alternatively, to record in modulation data – almost like an automation track within the architecture of the synth itself. To record an envelope into Sculpture, you'll

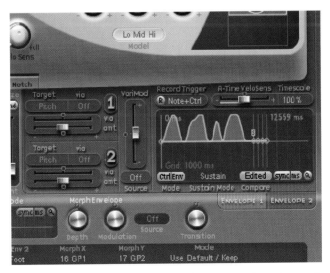

Placing the envelope into record mode will allow you to record your own "movements" into the envelope generator using the modulation wheel.

need to activate the small R in the top right-hand corner of the envelope display. This places the Envelope into record-ready mode, which, by default, can be initiated into recording by holding down a MIDI note and wiggling the modulation wheel accordingly. Press R again to stop the recording, and you should see your moves stored into Sculpture's envelope generator.

In essence, the Morph pad stores a series of five snapshots representing positions of the main Sculpture parameters (Object Strength, Inner Loss, Tension Mod, and so on), which can then be morphed using the central XY pad. The great thing here, from a modulation perspective, is that over 20 crucial parameters can be adjusted at the same time (indicated by the small red dots). Now add the fact that you can record these moves into an envelope to directly form part of the patch, and you have the world's ultimate vector synthesizer!

Arguably the first step to working with the Morph pad is to define a number of different parameter positions for each of the Morph pad's corners (A, B, C and D). Clicking on each respective corner, try adjusting the various key parameters, maybe applying differences in the Tension Mods, or the various positions of the objects on the string. The red dots provide reference to the current sound, also indicated by the position of the orange dot on the

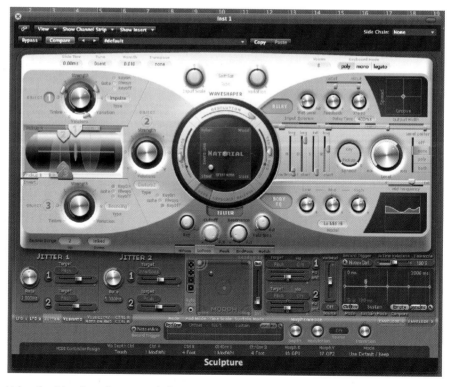

Using the Morph pad you can define positions for each of Sculptures key parameters, based on your respective XY positioning.

Morph pad itself. Note that you can also call up the Morph pad's contextual menu (by Ctrl-clicking), so as to copy and paste settings between the different points, or set the randomization destinations (initiated by the Rnd button) to have Sculpture "tweak" the various positions accordingly.

In some situations, you might find it adequate enough to have hands-on control with the Morph pad, and morph the patch as you see fit. Note that the MIDI controller assignment section to the bottom of the interface details the standard MIDI assignment with Sculpture, allowing you to map the pad's XY axis to two physical MIDI controls on your keyboard. Moving beyond that, though, you might want to record your own vector envelope – unique to the Morph pad – that will then be permanently embedded into the sound. This can be especially effective on pads – with each note taking on its own "wandering" quality as the sound slowly shifts through different phases.

To engage the vector envelope, you'll need to change the Morph pad's mode from Pad to Pad and Env. As with the envelopes in Sculpture, you can record your own moves directly into the synthesizer. In this case, enable the vector envelope Record Trigger (using the small R), play a MIDI note, and wiggle the Morph point accordingly to record the moves in. Pressing R again will stop the recording, with the moves displayed on the pad to illustrate the movement between the various vector points.

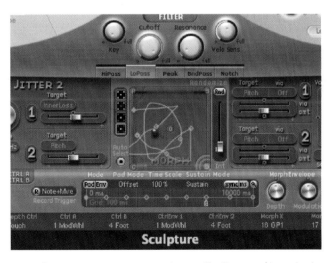

Recording your own vector envelope will allow you to embed Morph pad movements directly into your patch.

7.13 Creative sampling

Having explored some of the basics of using the EXS24 in Chapter 6, we now turn our attention to some of the more advanced creative aspects of using Logic's integral sampler. The assumption here is that you're using the EXS24 either to map samples (maybe from a sample CD or recorded into Logic), or alternatively, you want to explore some of the deeper mapping and synthesis options within the existing EXS24 instrument library. Either way, you'll find the EXS24 to be a highly-developed sound design tool, and an interesting way of dynamically developing audio files that you've recorded into Logic.

The EXS24 hierarchy: zones, groups and editors

As with any sampler, it's important to get to know the various hierarchical aspects of its operation – either in the form of different software pages or the way in which the sampler organizes and manages its sample data into zones, groups, and so on. One of the principal differentiations to make is between the main plug-in window – from which you can select instruments, apply envelopes, filters, and so on – and the EXS24 instrument editor window; which is specifically used for the mapping of samples across the keyboard. For most day-to-day activities, you'll only need to use the main plug-in window, however, if you intend to explore more creative applications of the EXS24 sampler, you'll need to make more extensive use of the EXS24 editor.

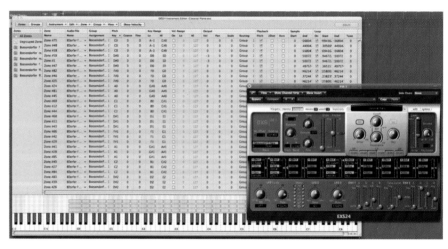

The two different sides of the EXS24 sampler – the main plug-in's interface window, and behind it, the EXS24 instrument editor where all the sample data is mapped.

Within the editor itself, Logic also makes an important distinction between zones and groups. A zone is smallest piece of currency in the EXS24 editor, which essentially holds a selected sample mapped across a corresponding range of keys – like C3-D3. A group is a handy way of organizing the various layers of sample data that can be included in an instrument. For example, a group could be used to differentiate between different velocity layers (soft and hard, for example), or two different components of a hybrid patch (like string samples in one group, and piano in another). There're also several interesting functional properties of groups – like the ability to apply different "conditional" triggering options, as well as the option to differentiate alternative filter settings, and so on.

7.14 The EXS24 instrument editor

The EXS24's sample editor can be opened using the small Edit tab in the top right-hand corner of the EXS24's main plug-in window – opening up the current selected instrument in the editor. In many ways, it's a good idea to try opening an existing library instrument first – both as means of getting a grip on how the Instrument Editor presents its mapping, and also to see how Logic's sound designers have gone about using the zone and group features to create their instruments.

The EXS24 editor is divided into a series of areas, allowing you to best visualize and navigate the mapping that you've created. First of all are the two main view tabs in the top right-hand corner – Zones and Groups. These allow you to quickly move between viewing the entirety of the sample data and its mapping – zones, in other words – and the "macro-level" properties of a patch organized into groups. For more rudimentary EXS24 patches, it might be that you only need to work in the zones view. However, as your programming gets more complicated, it might become easier to manage by viewing the instrument in both the zone and group view.

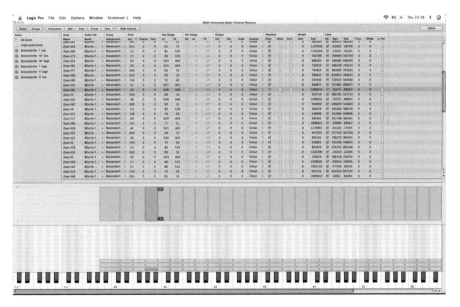

The Editor interface is organized into a series of areas, allowing you to visualize and edit the mapping of your samples in various ways.

In either view, the screen is divided into four key segments. At the top you'll find the parameters area, which provides a long list of the samples or groups included within the instrument, alongside their associated parameters (like key range, and so on) that define how they function. Moving down from this is the velocity area, which provides a quick snapshot of how samples are placed with respect to velocity – it might be, for example, that a zone or group works over the full MIDI velocity range (0–127) or only across a specific set of velocities (0–40, for example). Below this is the zone/group area that shows the placements with respect to the MIDI keyboard, which is viewable itself in the very bottom of the interface window.

Knowledgebase 7

ReCycle Files and the EXS24

Having seen how the Arrange area handles ReCycled files, it's also worth noting the distinct benefits and differences in working with REX files in the EXS24. Put simply, the reason for using the EXS24 over the Arrange page technique is creativity. As we'll see, once the REX files are imported into the EXS24 you can start to manipulate the raw data in any number of ways, as well as taking advantage of its tempo-tracking abilities by virtue of the loop being cut into a series of slices.

To import the Rex file you'll need to be in the EXS24 editor, selecting the ReCycle convert option from the Instrument menu. Arguably the best solution to pick is the Extract MIDI Region and Make New Instrument option, which will not only make a new EXS24 instrument from the audio contained in the REX file, but also place a MIDI region file into your arrangement to trigger the samples in the appropriate way.

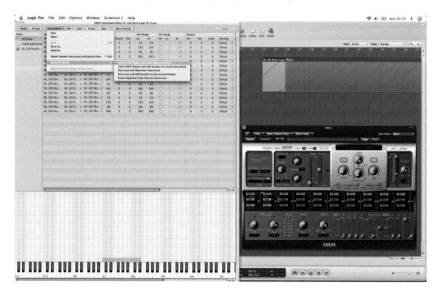

Importing REX files into the EXS24 is by far the most creative way of dealing with ReCycle files, facilitating a wealth of addition sound processing and re-sequencing options.

Once imported, the world is your oyster! Using the editor itself you could decide to experiment with some of the zone parameters, with some of the most distinctive treatments coming from the selective reversing of certain slices. On the front panel of the EXS24, you could also decide to experiment with some of the synthesis features, maybe using the filter to make the loop darker, for example, or changing the envelope settings – maybe using a sharper decay setting with no sustain to produce a chop- pier output. Looking at the region itself, you could decide to re-sequence certain elements (removing or retriggering certain slices, for example) or even experimenting with different types of quantizing.

7.15 Creating a new instrument and importing samples

Unless you're interested in remapping existing sample data, you'll need to create a new, blank instrument to work with. You can do this via the EXS24's Instrument Editor using the menu option Instrument > New.

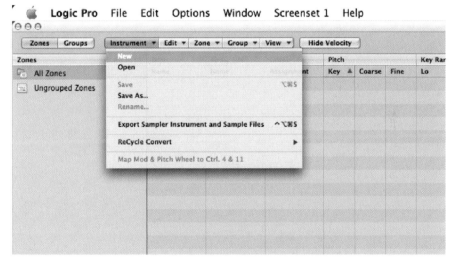

To build a sample mapping from scratch, select the New option from the Instrument menu.

Before you start to import sample data, you might want to consider some essential data management issues. For example, if you intend to keep your sample data in a unified place (like an external FireWire drive, or a separate partition) you'll want to copy the sample data into an appropriate folder first, before initiating any imports into the EXS24. Of course, it is possible for the EXS24 to reference sample data from any drive – so if you're feeling lazy, and have no intention of keeping your data in a unified place, just go on ahead and import! Finally, it's also possible to store sample data as part of your project's assets, which might at least facilitate a backup of the sample data, although there's no guarantee the backup will exist on the drive 3 months down the line!

The importing process itself can be done in a two principal ways – either via the Zone menu (usually when you intend to load multiple samples in one action) or using a drag-and-drop from the project's Audio Bin and/or the Browser. If you've invested the time in naming your samples properly, or indeed, chosen to import samples from a third-party collection, then the Zone > Load Multiple Samples is probably the quickest and easiest solution. From this option you'll be able specify the samples to be loaded, followed by a dialogue to provide some indication how these should be mapped. The options are relatively self-descriptive, but see the following text for clarity:

Use Auto map for a collection of keyboard samples where the samples span more than one key. In this case, Logic will map the sample intelligently (using the root key included in the samples' name) running each zone into (but not over) the next.

The Drums option again auto maps the samples based on the root key in the sample's name, although this will be the one and only key the sample is mapped to.

The Contiguous zones is useful where you don't have key information included with the samples, or simply want to ignore any existing key information included with them. Using the Contiguous zones option you'll also need to specify the zone width for each zone (this could be one note, or a whole octave or more) as well as the start note to begin the mapping.

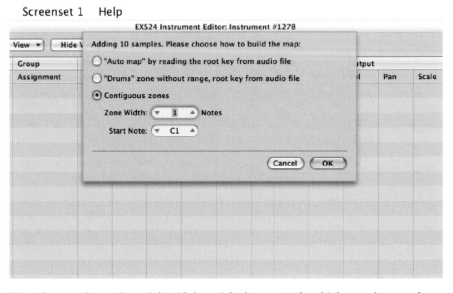

Using the mapping option might aid the quick placement of multiple samples – say from a sample CD.

To build the instrument mapping up in a more gradual way, use the Browser window and drag-and-drop the files as you see fit. The exact way the samples are mapped will vary as to the samples selected and how you drag them into the EXS24 instrument editor. Dragging a single sample directly into the zone parameters area, for example, will place the sample across the entire keyboard with its default pitch at C3 (unless specified in the file name). Alternatively, drag a sample onto the MIDI keyboard on the bottom of the editor window, and you'll be able to map it to a specific key, although the range won't extend beyond that single note. Dragging onto an existing zone will also preserve the current zone parameters, but use the newer choice of sample.

Using the browser, you can also select multiple samples and import these directly into your mapping. As with the dialogue from the Zone menu, you can also specify how Logic auto maps the samples. Note that by dragging the multiple samples onto a key on the MIDI keyboard, you'll effectively set the starting point, with subsequent zones following on from this point.

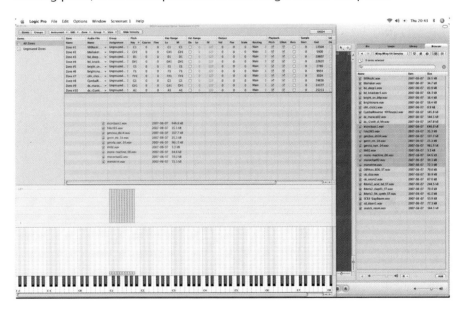

As well as using the Import option from the Zone menu, you can also drag and drop samples into your EXS24 instrument directly from the browser.

7.16 Changing zone properties

As part of the importing process, it's highly likely that Logic will have performed some basic zone parameter adjustments. Let's take a look at the different sections in the zone parameters areas and how you might go about using them.

Key range

As we've already seen with the process of drag-and-drop sample assignment, each zone is assigned its own unique key range, with both upper and lower settings. If you're creating a drum-based instrument, it's likely that the Lo and Hi setting will be on the same note. However, if you're assembling a multi-sampled instrument – maybe with two or three samples for every "played" octave – you'll need to adjust the range over a few consecutive notes. You can adjust the Key range via the zone parameters area, or indeed, by dragging the appropriate "mapping blob" in the bottom zones/groups area.

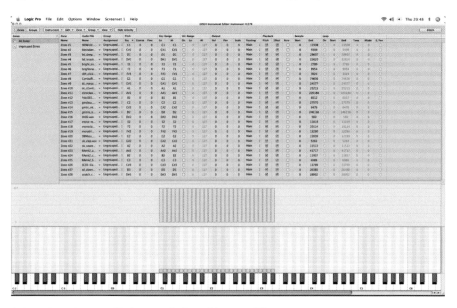

A drum-based map will place each drum hit on a different note.

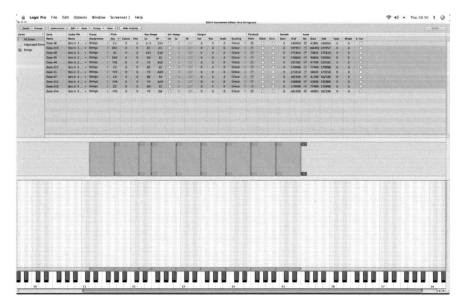

A keyboard-based multi-sample might use one sample to cover two or three notes – in this case, the key range should be adjusted accordingly.

Pitch

Pitch defines the fundamental root position of the note – in other words, the note on the keyboard that plays the sample at its natural sampled pitch. In the above example of a multi-sample spread over a few notes, you'll probably want to place the key somewhere in the middle of the zone's range, so as to keep the amount of transposition at a minimum. You can also use the Coarse and Fine tuning parameters to fine-tune the pitch of the sample – maybe the instrument has been recorded slightly out of tune, for example, or you want to explore some more creative aspect of tuning on drum and percussion samples.

Vel. Range

Zones can be set to work over a specific range of velocities, using the Vel. Range parameter. If you've got several samples taken at different velocities, this might be a useful way of achieving a realistic sense of expression in your instrument, although you can always "fake" velocity sensitivity by some careful use of the amplifier and filter responding to velocity.

For more complicated patches – where you might be dealing with several different velocity layers – consider applying the velocity switching via the group feature.

Output

Output allows you to fine-tune some important mix-related issues with regards to your zone. Volume and Pan are largely self-explanatory, but what's of particular interest is the Routing option. By default, each zone is set to the main outputs, but it's also possible to force a zone out to any of the EXS24's 16 individual outputs. To access this, though, you will need to instantiate a Multi Output version of the EXS24 in your instrument channel, and activate the additional Aux channels accordingly.

By routing a zone out in this way, you'll be able to gain individual access to audio plug-ins as well a discrete channel fader for the zone(s) in question. This is particularly useful editing tool for drum-based instruments in the EXS24 library that you might want to enhance with additional processing, like compression, equalization or reverb.

Playback

Playback governs some important aspects about how the sample data is read, mainly in relation to triggering drum samples. For example, the Pitch parameter allows you to enable or disable pitch tracking with a sample, so that even if a drum sample spans several keys, for example, its pitch remains fixed. 1Shot, on the other hand, focuses on the duration of the sample, so that in 1Shot mode the full length of the sample is played irrespective of how long the key is held down for. Again, the main application of 1Shot mode is with drums, where the duration of a hit isn't "variable" as such – in other words, a cymbal will always decay over a number of seconds.

Rvrs (or reverse) allows you to flip round the playback of the sample data. Given the need to create new audio files before you can reverse region in the Arrange Area, this EXS24 reverse function is a quick-and-easy solution in creating reverse effects. Needless to say, the reverse feature is a great creative tool, sounding great on a range of instruments – from drums hits, to pianos and bells.

Sample

Although most samples that you'll import will have their edits already in place, it is possible to further refine the start and end point from within the EXS24 Instrument editor. For example, it might be the case that you have a small portion of silence at the front of the sample, or that you want to create a more abrupt end to the sample. Although the EXS24 Instrument editor displays this numerically, you can Ctrl-click on the sample number to open the sound file in Logic's Sample Editor. If you're running the editor at full-screen size, though, you'll move the EXS24 editor out the way to reveal the Sample Editor.

Loop

The Looping feature is primarily used as means of creating an infinite sustain on sampled instruments like strings, and so on. Obviously, if a sample was taken for the full-note duration, this could make an exhaustive drain on RAM allocation (although some third-party libraries do this), so samples are taken of about five seconds or so in length, with the last three seconds looped over until the notes end.

In the case of the EXS24, you can define a number of parameters in relation to loop, with the principal aim of creating a smooth, glitch-free loop that

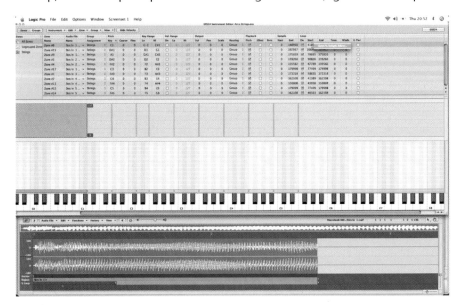

Although it is possible to set the loop numerically, opening the loop points in the sample editor (by Ctrl-clicking them) will allow you to better place the start and end points to match what is happening in the waveform.

shouldn't be particularly audible to the ear. The beginning point is always defining the Start and End point of the loop – ideally picking a section of the sample that doesn't waver or drift too much, but instead, has a clearly defined sustain. Although you can define a loop point numerically, the easiest way to do this is to open up the Sample Editor (by Ctrl-clicking on either the start or end parameter number) and perform this visually. Make sure that you've got the Search Zero Crossings enabled (under the Edit menu), so that any start and end points snap to a zero crossing point.

Even if you do find a relatively successful location to position the start and end points, it might still be important to use some of the EXS24's additional loop parameters to fine-tune the looping, and so produce an even smoother looping effect. Xfade, for example, can be useful if you're still hearing small clicks or jumps around the loop point, with a larger number creating a longer crossfade. E. Pwr further refines the crossfade, so that no energy is lost in the crossfade processing, while tune addresses any tuning issues that might have become evident in the looping process.

7.17 Working with the EXS24's groups

As soon as you start building up instrument files with more than 10 or so zones, it becomes essential to start to manage the samples in some meaningful way. Groups offer a variety of benefits to the EXS24 sample list – from

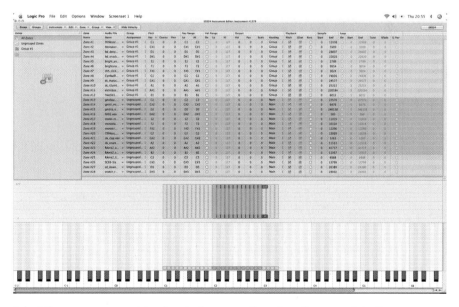

Select the zones you want to place in the group then drag them over into the zone column to create a corresponding group.

223

organizing better ways of displaying related zones, to advanced triggering options. Using the instrument editor you can create groups in much the same way as you might create playlist in iTunes – simply select the zone or zones you want to group and then drag these across to the zone column on the left.

In a rudimentary way, the zone column offers an excellent way of navigating your instrument – either viewing the complete set of zones in an instrument, or choosing to view the instrument group-by-group by clicking on the appropriate group folders in the zone column. Note that you can also name the groups by double-clicking on the group name. This is also an excellent way of organizing and creating hybrid EXS instrument – selecting different layers of zones from one instrument and copying and posting them into another.

Advanced mapping options

Moving over to the groups view in the EXS24 instrument editor, you should also see a number of parameter options unique to groups you have created. What groups offer is a form of macro-level control over an instrument – so, rather than having to individually set velocity switching over hundreds of zones, we can place them into four respective groups and then simply alter the velocity setting of those individual groups. Not only are grouped velocity adjustments far quicker in the short term, but also for any adjustments later on, the group feature is invaluable.

Groups contain a unique series of mapping options – including filter and envelope offsets and various conditional triggering options – ideal for the expert EXS24 user.

As you'd expect, the Key Range, Vel. Range and Output parameters function in much the same way as they did in the zone parameters, although of course, settings here override above what you've established in the zone (for example, if the group's key range doesn't extend as far as where you placed a zone, you won't hear the sample).

The Trigger option defines whether a group is triggered by a Note On or a Note Off command (known as Key Down and Key Release, respectively) – an effect more broadly referred to as Release triggering. Release triggering uses short samples of events that occur after the main note to create a more realistic and lifelike "musical event." Release samples could be in the form of reverb, for example, or the small decaying "resonance" when a piano note is released.

Following on from the trigger options – although positioned over the far right-hand side of the group parameter – are the conditional Select Group By options. Although you can create a number of different conditional triggering options here, probably the most well known is the option to create a so-called key switched instrument. With key switching, a number of notes down the bottom of the keyboard will be used to dynamically toggle between different instrument articulations (like staccato and legato), all on the fly. Each articulation, as such, is organized into its own group, with the group "selected by" a specific note, like C1.

The Select Group By actually works with a number of different MIDI events (Note, Controller, Pitch Bend and Channel). This makes conditional triggering an incredibly flexible feature – allowing you to use a modulation wheel, for example, to switch between layers.

The Envelope and Filter Offsets allow you to differentiate between the various groups and the initial settings on the EXS24's front panel (which we'll look at in more detail in a minute). For example, although you might set a master filter cutoff position at 50%, for example, you could use the filter offset parameter so that one group opens the filter in a more open position (+50), while the other closes in down (−25). Although this isn't quite the provision of a completely independent multimode filter per group, it does allow different zones to use different filter settings concurrently – in other words, while one group sounds filtered, the other is less so.

Following this concept further are the envelope offsets, which again work from the basic premise of the parameters established on the main front panel, but then adapted, negatively or positively, to match the specific needs of the group. You could, for example, have a nylon string guitar in one group using the fast attack and release settings, while in another group, some string samples use softer attack and release settings to produce a lush background pad.

7.18 Editing EXS24 instruments using the front panel

With the samples mapped, now comes the task of adjusting and refining the synthesis engine of the EXS24's sample playback. The settings on the front panel extend what you've already achieved in the mapping of the samples. For a simple instrument, for example, all you might need to adjust is a few key parameters like release, or its velocity scaling. Alternatively, you could use the front panel of the EXS24 to completely transform the samples way beyond their original sound – maybe using some extreme filter and drive settings, combined with a series of creative modulation matrix settings.

Like the ES2, the front panel is divided into a number of key sections, including the filter block, the modulation matrix and the LFO and Envelope modulators. To understand how these parameters work, lets look at couple of different editing scenarios and how these might work:

Basic instrument refinement: velocity scaling, polyphony and envelopes

Not all instruments require a great deal of modification away from the default settings that appear on the EXS24's front panel, but there are a couple of areas worth looking at to best optimize the instrument in the final mix. First of these has to be the volume controls over the right-hand side of the plug-in interface. Alongside the overall volume level (which might be worth reducing if you're using a lot of sample voices simultaneously), you'll find the sampler's scaling of level via velocity. Raising the bottom part of the level via velocity slider can sometimes be a useful alternative to compression, especially if you've got a multisampled instrument with "quieter" samples in its low velocities.

Adjusting Envelope 2 – which, by default, is assigned to the Amplifier duties – is a useful way of quickly grabbing hold of, and shaping, an instrument's ADSR characteristics. For example, if some of the samples in your instrument have reverb embedded into them, you might find it beneficial to increase the release time so that you hear the full decay of the room ambience. Alternatively, using a slower attack can be a valid sound design treatment to soften or disguise the initial attack transient of a sample.

Towards the top left-hand corner of the interface are a number of useful macro controls, some of which can have a big impact over the performance of the sampler. For example, the voices parameter reserves a given amount of polyphony for the instrument in question. Sixteen notes, for example, might be more than enough for a simple drum-based instrument, but should you use something like a multi-sampled piano with release samples, you might need to raise this up to the full 64 voices. However, with an increased voice count comes an increased CPU drain, so it's often the case that you actively try to balance out the realistic demands of the instrument, against the overall needs and drain of your CPU performance. A Used indicator at least provides some rough indication as to the total used polyphony, but you should also be able to hear a "maxed-out" polyphony by longer notes appearing "snatched off".

Adding in the filter

As we saw with the ES2, the filter is a surprisingly powerful way of shaping the timbre of our sound source. The filter included in the EXS24 is certainly no exception to this rule: with up to six different modes of operation (including high-pass, band-pass and four different strengths of low-pass), some fat and juicy Resonance, and even a drive parameter.

To activate the Filter you'll need to press the on switch, otherwise it defaults to its off position so as to save a few CPU cycles. Switching the modes can be done using the six tabs at the bottom of the filter, with the additional Fat button to preserve some of the low-end with high-resonance settings on the

The main controls in relation to volume are on the right-hand side of the EXS24's interface. Try experimenting with the sampler's response to velocity, as well as how Envelope 2 changes the amplitude over time.

Once you've engaged the filter, you can use its variety of modes (low-pass, band-pass, and high-pass) to shape the timbre of the EXS24's output.

LPFs. Remember that the precise position of the filter cutoff and resonance can also be varied between different groups, using this front panel setting as the initial starting point.

Modulation heaven!

Many of the more advanced synthesis options come via the application of the modulation matrix, which allow you to map a variety of sources (including the LFOs and envelopes at the bottom of the interface) to various destinations within the EXS24's synthesis and sampling engine. As with the ES2, the EXS24 uses a familiar matrix display, with a Destination parameter (like Filter Cutoff), and amount of modulation and the Modulation Source including any of the EXS24 modulators (three LFOs and two envelope generators) or any one of the 127 standard MIDI controllers.

As with the ES2, a large Modulation Matrix section allows you to route a number of different modulations sources to targets in the EXS24's programming architecture.

Saving instrument settings

Confusingly, the EXS24 makes some level of distinction between the instrument settings stored with and without the front panel controls. To permanently store the front panel setting alongside your instrument mapping you'll need to go to the Options menu on the EXS24 and select the option "Save settings to instrument." Now, when you return to that instrument, the mapping will be loaded alongside any front panel settings that you've created. Alternatively, use the

If the settings on the EXS24's front panel have become intrinsic to the instrument in question, consider using the Save settings to instrument feature under the options menu.

Recall setting from instrument or Recall default EXS24 settings if you've made changes to the front panel settings that you're not happy with.

7.19 Ultrabeat

Ultrabeat is Logic's take on a contemporary drum machine, complete with a range of synthesis and sequencing options, as well as a handy multichannel version to allow the flexible access to Logic's signal processing features. An Ultrabeat Kit is made up of up to 25 different drum voices – listed down the left-hand side of the interface complete with level controls (as the blue bars), mute, solo, pan and an output selector. On selecting a drum voice you can then edit its associated synthesis parameters in the large central display, or indeed, program its unique step sequence at the bottom of the plug-in.

Ultrabeat features up to 25 different drum voices in each kit, each with their own set of synthesis options.

In keeping with flexibility of the ES2 and Sculpture, the synthesis options in Ultrabeat are impressive to say the least! Each drum voice has access to up to three different oscillators, which are then passed through a multimode filter a ring modulator, and finally some EQ. Arguably the most useful oscillator, though, is oscillator 2, which features a sample playback mode as well as the

229

phase Osc and model modes. On the whole, you'll find that samples offer the most quick and effective solution for assembling a kit (if so, use the "blank" Drag & Drop Samples Kit from Ultrabeat's plug-in settings), although it's also good to augment these with some of the pure synthesis elements, like the noise generator. Once the sample mode is active, you can also drag sample files directly from the browser onto oscillator 2.

Use the Browser to drag and drop samples directly into Ultrabeat's second oscillator.

Clicking on the filter tab activates the filter, offering a great way of colouring the samples, like a band-pass filter applied across a clap sample to make it sound narrower. To modulate the filter, you can also use any one of the four envelopes and an LFO, with the small blue legend beneath the cutoff and resonance controls allowing you to define an appropriate modulation source. If you're after grit, though, try activating either the crush or distort modes towards the bottom of the filter dial – both effects that can sound great on heavier kick or snare samples that you want to push to the front of the mix.

When it comes to assembling the sequence, you can just step through each voice and create the finished pattern piece-by-piece. A much more effective solution, though, is to switch Ultrabeat into its Full View mode using the small tab in the bottom left-hand corner of the interface. In full mode, the synthesis controls are removed to reveal a full-sized drum grid for all of Ultrabeat's 25 drum voices. This makes it far easier to gauge the interaction between the various parts, rather than having to remember their various hit points. Note that to hear the sequencer in action, you'll need to activate its power tab, with the transport controlled either from Ultrabeat itself (in which case it plays independent of the rest of the project) or from Logic's main transport, so as to audition the pattern in the context of the song.

In total, you can sequence up to 24 different patterns with Ultrabeat, with the potential to switch between them using a MIDI keyboard. A much better solution, though, and one that's makes more sense in the Arrange Area, is to export the pattern out as a region. To do this, click-and-drag on the small icon next to the pattern selector, placing the resultant region on Ultrabeat track in the Arrange area. Exporting a number of patterns in this way, you can then revert back to Arrange area to structure your song accordingly, even using other MIDI editors (like Piano Roll or Hyper Editor) to further manipulate the data.

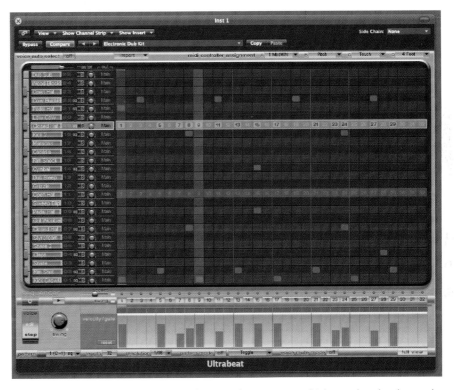

Use Ultrabeat's Full View to sequence the complete pattern, which can then be dragged directly into Logic's Arrange Area.

Mixing in Logic

8.1 Introduction

Up until a few years ago there was only one accepted way of creating a professional sounding mix – hiring a commercial studio facility and mixing through a traditional mixing console, like an SSL or a Neve. However, times have changed significantly since then, and mixing "in the box" – in other words, mixing completely within the domain of your audio sequencer – has become a viable, and some would say more flexible, alternative to the traditional console route. In truth, applications like Logic offer a tremendous amount of creative and technical freedom to today's mix engineer: complete and instantaneous recall, compression and equalization on every channel, studio-grade reverb, full automation and much more besides.

In this chapter, we're going to take a closer look at the process of mixing in Logic, looking both at the technology it has to offer to create an effective mix, and how we can knit these elements together as part of the mixing workflow. This is also an opportunity to get to know many of Logic's essential signal processing plug-ins in a more informed way, as well as taking a more detailed look at how the Mixer area works – aspects that can actually improve the entirety of your workflow in Logic, as much as the mix. Looking more closely at the process of mixing itself, we'll explore how to use compression and equalization to create separation, as well a bus processing (on aux channels), and of course, spatial treatments like reverb and delay that help define the depth and dimension of your mix.

Beyond the essential signal processing tools of mixing, we'll also take a look at features like automation, grouping and folders, all of which help make your mix easier to manage, especially with the large track counts that most projects seem to encompass.

8.2 Channel strips: understanding your virtual console

Before diving headlong into your first mix on Logic, it's worth taking some time to distinguish between the various channel strips that the mixer is comprised of, and how these various elements interact with each other. Generally, a good mix in Logic will make use of all the features of the mixer – channel inserts, bus sends to aux channel strips, output masters, and so on – so it's worth familiarizing yourself with their features in much the same way as an engineer would "get to know" the sections of a physical console.

Audio channel strip

An audio channel strip.

An audio channel, as we've already seen in previous chapters, governs the basic path in and out of Logic for a recorded signal. In a mix, audio regions in the corresponding track will be sent down the channel strip, through the various insert processors (compression and EQ, for example) and bus sends (for reverb and other effects) to a designated output.

Instrument channel strip

The instrument channel strip duplicates the same features as an audio channel strip; only this time you get to work with the signals generated by the virtual instruments in your session. In addition to this simple application, an instrument channel strip can also be configured for Multi Channel operation so that multiple outputs from the same instrument (like the EXS24 or Ultrabeat) are sent to a number of additional auxiliary output channels. Besides having individual level control, this also allows you to apply different effects onto each output so that a snare, for example, might have different equalization, compression and reverb to that of a kick drum.

An instrument channel strip.

Plug-in boxout 1

Distortion Effects

Logic includes an impressive set of effects dedicated to low-fi transformations and crunching-up sounds – from the digital extremes of Bitcrusher, to the warmer tones possible with Guitar Amp Pro. Besides being an obvious addition to guitar parts, distortion can also be a great source of colour and interest on any number of sounds on a mix. Try using distortion to add some grit and body to a drum loop, for example, or a touch of drive on an aggressive vocal.

The first choice of amp-like distortion would have to be Guitar Amp Pro, which emulates the Amp, EQ and speaker components of a typical guitarist's amplification system, as well as modelling a number of different mic'ing options. Interestingly, many guitarists are choosing to record their guitars DI'd directly into Logic, and solely use Guitar Amp Pro as a means of adding the required amount of distortion and speaker colouration. The clear advantage here is that the tone of the guitar can be completely modified (changing the Amp head, for example, or the position of the virtual microphones) at any point in the production process, without having to re-record the guitar. However, to get the full range of tones possible from a guitar rig you may have to use some additional plug-ins – for example, try adding in a compressor (ahead of the amp) to create more sustain.

Distortion, overdrive and clip distortion come from earlier incarnations of the application, and tend to be less suitable for applications on electric guitar. They are, however, an excellent addition to drum loops or acidic synth lines generated from the ES2. Both distortion and overdrive have particularly simple controls with just a drive and tone parameter, alongside a corresponding output reduction slider to avoid ripping your monitors to shreds!

The prize for the most bizarre distortion effects goes to the Phase distortion and Bitcrusher plug-ins. Phase distortion sounds great on drums loops needing to be mashed up beyond all recognition. Based on a modulated-delay line, it can produce tones that range from a warm fuzz to the sound of a FM broadcast gone considerably wrong! Bitcrusher, on the other hand, is a great source of digital grit – especially as you reduce down the bit slider (to around 8 bits) and start applying some heavy Downsampling ($10\times$ and beyond).

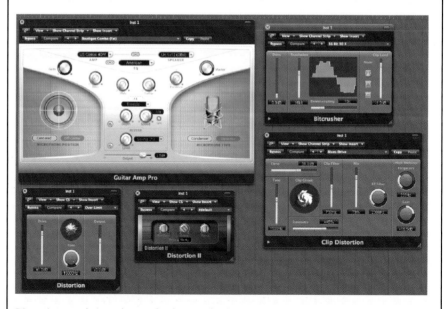

Distortion can bring plenty of colour and grit to your mix. Try experimenting with Logic's wide range of distortion effects to achieve anything from subtle valve-like warmth to edgy digital distortion.

Aux channel strip

An aux channel strip.

The humble aux channel strip is one of the most useful parts of Logic's mixer, and the source of a number of different techniques in mixing. Firstly, they can be used as means of applying send effects like reverb or delay – anything where you want to actively control the balance of wet and dry sound using a separate fader. In this example, the aux channel strip has the effect strapped across its insert path with instrument channel strips or audio channel strips sending signals via designated bus sends. This allows any number of channels to access the same reverb, for example (an essential way of preserving DSP resources), as well as being able to control the return of the effect in the same way as any other audio signal in the mix – in other words, it could be equalized, compressed and controlled by movements of the fader.

Another less immediate, but altogether just as useful, technique is to use an aux channel strip to combine a number of different channels (including audio and instrument channels) on a single fader. For example, many engineers using a traditional console will create a sub-mix of the drums to a selected bus master fader (in this case, one of Logic's aux channel strips), allowing them to quickly control the level of the drums relative to other instruments in the mix. Additionally, they could also make use of inserts on the bus faders to apply compression, EQ, and so on, to the entirety of the drums, rather than individual channels within it – something which can just as easily be done of Logic's aux channel strips.

237

The final important role of the aux channel strip is to operate as a means of inputting external signals into the mix. If your audio interface supports enough inputs, there's no reason why you can't patch-in external compressors, equalizers, synthesizers, and so on directly into your mix. Although you might need to account for a small amount of latency, using dedicated recording hardware can often supply much more character than conventional plug-ins. Of course, given the provision of inserts on these aux inputs, there's no reason why you can't also use Logic plug-ins on top of whatever device you're inputting into that particular aux input.

Output channel strip

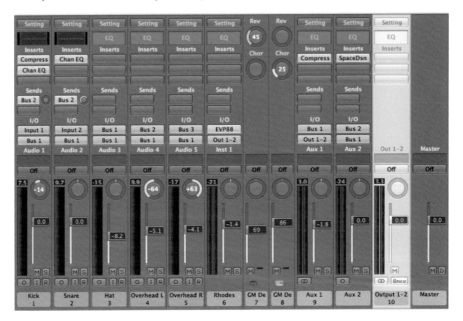

An output channel strip.

The output channel strip represents each of the physical outputs in your Logic system. If you're using a simple two-in, two-out USB audio interface, you'll only see one of these faders. However, if you're working with a multiple output FireWire or PCI soundcard, you'll see a number of software outputs corresponding to each physical output. On the whole, most mixes will simply use one designated output (1–2) as the "destination" of the two-track mix. If you intend to render the mix directly from the Logic session, you'll also need to make use of Bounce button, found only on the output channel strip (for more information on bouncing to disk see Chapter 9).

Plug-in boxout 2

To Gate or Not To Gate … That is the Question.

Music trends change, so it's not a surprise that the process of recording also adapts alongside. For example, just a few years ago it was common-place to gate the components of a drum mix (kick, snare, toms, and so on) to within an inch of their existence! Nowadays, it seems that listeners, engineers, and musicians all seem to prefer a looser sounding kit, resulting in much less use of gating.

Now and again though, you may well feel the need to improve your aural hygiene – maybe a vocal mic has some obtrusive background noise made worse through the application of compression, or that one and only guitar take has a little too much noise from the amp. If you're pushed for time, there's no doubt that the noise gate plug-in remains a quick and effective way to attenuate problematic noise in between notes – although remember to gate before the compressor and not after it! The trick with setting the noise gate is to start with hard settings (quick Attack, Hold and Release, with 100 dB of reduction) to find the right threshold. Although the gating will sound harsh, you'll be better able to find the "sweet spot" just above the amplitude of the noise. With this set, back off some of the settings, especially Release and Reduction to soften the effect – surprisingly, even small amount of reduction (6–10 dB) can have a big effect on the overall cleanliness.

If you've got a little more time, most users now tend to approach the gating issue through a few crafty edits. Try using the arrange window's Strip Silence option (Audio > Strip Silence) as a means of Logic preparing the majority of edits for you. As with a conventional noise gate, you'll need to establish the right threshold to get the best results, although arguably this is easier to "see" on Strip Silence than it is to "hear" on the noise gate. However, the big advantage with the Strip Silence approach is that you can modify the edits later on, making sure none of the important attack transients are missed out.

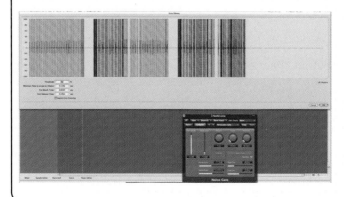

Two choices for improved aural hygiene – the traditional noise gate, or the current favourite, Strip Silence.

239

MIDI channel strips

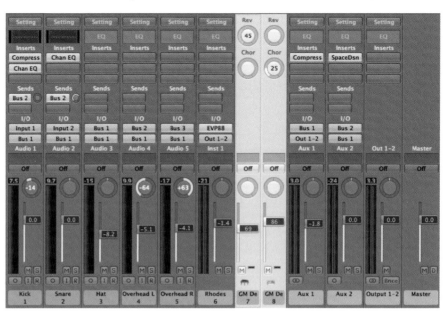

A MIDI channel strip.

Arguably, MIDI channel strips should be kept distinct and separate from the audio mixing, but given their inclusion as part of the mixer it's worth clarifying their exact role. Unlike all the other mixer channels in Logic (which control the flow and qualities of audio within the application), the MIDI channel strips have no direct control of the "internal" properties of your mix, but instead work as controllers for external MIDI hardware connected to your MIDI interface. However, in the case of the synths and sampler being returned via aux channel strip, these MIDI channel strips could have a direct effect on your audio mix – maybe balancing 16 MIDI channels being returned to one stereo input on your audio interface.

Master channel strip

The master channel strip provides a quick-and-easy access point to the level of signal going to all main outputs, and is directly linked to the "volume" control as part of the features on Logic's transport bar. In situations without control room monitor levels, for example, this could be used as a means of adjusting the monitoring level, although it should be noted that any digital bounces made in Logic will be subject to the master channel strip's gain adjustments.

Note that you cannot insert plug-ins across the master fader in stereo mixing (instead use the output channel strips), with the master channel simply being used a level attenuator. In surround mixing, however, its role – and the role of the outputs – slightly changes, which we'll cover in more detail in Chapter 10.

8.3 Organizing your mixer: what you do and don't see

Ask any good mix engineer about the fundamental tool that helps them negotiate an effective mix workflow and they'll often reply "a well-organized mixer." On a traditional console, for example, an engineer might decide to re-patch the tape returns so that the order and arrangement of the various channels best reflects how they might want to carry out the mix – maybe with the drums in the first 8–10 channels, for example, while bass, guitars and keyboards following on subsequent channels. So, before you start diving into equalization settings and designing your soundstage, take some time to organize both the arrangement and track list to help navigate your way around the mix successfully. Logic also provides a number of ways to organize and structure the appearance of your mix: separating off sections, for example, hiding unwanted channels, viewing the complete signal path, and so on.

Mixer views

The first concept to grasp with respect to using the mixer interface effectively is the link between the Arrange window and the Mixer window. By default, the mixer is set to its Arrange view out of the three possible view modes at the top of the Mixer window – Single, Arrange and All. In Arrange mode, the order and amount of channels displayed directly corresponds to the Arrange track list. For example, you might have reorganized the vocals to place them on the first track lane – a decision that will also impact on the arrangement of the mixer. So, one simple way of organizing your mixer is to organize your track list!

Following this concept, you'll notice that the track Hide feature that we first saw in the editing chapter also carries through into the mixer. One intriguing difference to the behaviour in Arrange window, though, is that the Hide option has an immediate effect on the currently viewed channels, irrespective of whether the H button in the top of the Arrange area is active or not.

An alternative to the Arrange view is the All mode, which will display the entirety of audio channels in your Logic project arranged sequentially by their assignment (Inst 1, Inst 2, and so on), rather than their order in the Arrange area's track list. This is a useful way of seeing the entirety of your mix, including objects (like the metronome and Prelisten channel) that you wouldn't usually have access to. However, without the ability to organize this view, it is easy to become disorientated, especially in bigger mixing sessions.

In Arrange view mode, the Mixer area directly corresponds to the order or channels in the track list. The Hide track option can also be a useful way of removing further channels from view.

Knowledgebase 1

Clipping Faders: Good or Bad?

On the whole, conventional engineering wisdom tells us to avoid situations where a signal "clips" the meters – in other words, whenever we see one of Logic's channels faders or master outputs go into the red. However, in reality, there are some situations in Logic where this is acceptable, and others where it demands your immediate attention!

In Logic a fader will register a clip whenever it is presented with too much signal – this might be because you've brought up the output gain of compressor too much, or that you're feeding too much signal to the main output channels. Looking first at channel faders and instrument tracks, the clipping we see here isn't particularly problematic, with no immediate chance of you hearing distortion at first. This is all because Logic's mixer incorporates a degree of safety margin – or headroom – built in, using an enhanced bit-depth resolution to tolerate peaks in the region of +1 to +6 dB.

Moving onto the main output channel strip, however, things are not quite so flexible. Any final digital medium has a fixed amount of level it can tolerate – in other words, a CD or WAV file cannot exceed 0 dBFS (0 dB on the output fader) without the top of the waveform being clipped or distorted. In this respect, it is important to ensure that the main output fader doesn't ever go into the red when you print off your final mix (although the odd peak during the mixing process isn't too disastrous). Given the amount of signals potentially feeding this channel, though, this is quite easy to encounter.

So, if your output channel does clip, what should you do to resolve this? The first and easiest solution might be to turn down the faders contributing to the mix, which could be done by placing them all into a group, and then turning the group down accordingly. If it's in the region of a few decibels, you could also choose to turn the master channel strip fader down, although if you're using it in the realms of −6 to −10 dB then it suggests that there is a fundamental level mismatch that needs to be addressed. Another option is also to put an instance of the gainer plug-in (on the master output), and also use this to reduce to final mix level ahead of it being bounced.

Although clips on channel faders aren't too problematic, it is advisable to avoid clipping the main output.

Viewing by signal path

One of the most interesting view modes is the so-called Single mode, which shows the complete signal path of a selected channel. The signal path illustrates the full journey a channel makes through Logic's mixer, including all the associated bus send effects, any sub-mixing via aux channel faders, and of course, the final output channel strip. Using Single mode is a great way of honing in on the specific signal-processing characteristic of one sound in the mix, especially where it's using multiple bus sends (for reverb, and so on) that could be positioned well along the mixer in the standard arrangement view.

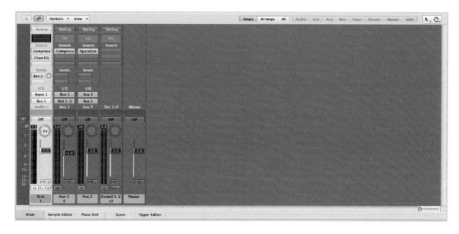

In Single mode you can see the complete signal path for the currently selected track.

Dropping out sections of your mixer

Using the various tabs along the top right-hand corner of the Mixer area, you can switch out various corresponding sections of your mixer. For example, you might choose to remove MIDI faders given that they don't directly input on the signal processing aspects of the mix. Alternatively, by removing the audio and instrument (Inst) from the view you can gain quick access to the aux masters and main output without having to scroll across the full length of the mixer.

8.4 Folders and the mixer

As we saw in Chapter 4, folders can be useful way of packing your arrangement into groups of related tracks, which can then be viewed and edited independently of the main arrangement. For example, you might decide to pack away your drum tracks into a individual folder, keeping those tracks away from the main view of the track.

Not surprisingly therefore, the folder feature also has an interesting impact on your view of the mixer. Any folder tracks are displayed as small strips in the mixer, without any form of level control, mute switching, and so on. However, by clicking on the small folder, or entering the folder via the Arrange page, you can access this sectioned off part of the mix, using the small hierarchy tab in the top left-hand corner of the mixer to move backwards into the main mixer.

Managed carefully, folders can be a useful visual tool for keeping on track of the sections of a mix; although, as we'll see later on, there are also other functions for the control of the groups of instrumentation from a sonic perspective. One of the tricks is to decide what you keep on the top level of the Arrange area. For example, if you delete the track lanes from the top "arrangement" level, they'll only be viewable once you've moved into the folder. This can be a great solution for hiding parts of the mix away, although not so good if you still want to keep an eye on the entirety of instrumentation. As an alternative, retain the track lanes in the arrangement (even though the regions have been packed away into a folder), only using the folder functionality to "hone in" on specific parts of the mix as and when you want to.

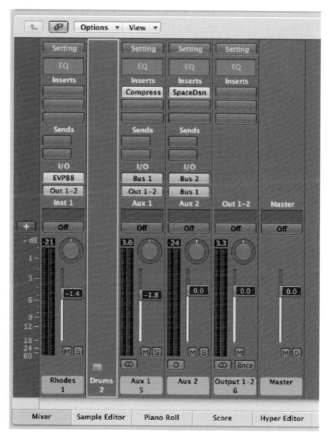

Use the folder function to "pack away" parts of the mix. This can be useful in projects with large amounts of channels or tracks.

Plug-in boxout 3

Helpers ... Handy Little Plug-Ins

Sometimes the small, innocuous little plug-ins can be vital to creating an effective mix as the big processor-hungry reverbs. Take the Gain plug-in, for example, its main application is simply to raise the gain of an input – surely not that useful then? Say you've recorded in a series of complicated automation moves, but simply want to raise the entirety of the track by 2 dB. Usually this would necessitate editing or re-recording the moves. Alternatively, you could use the Gain plug-in (patched somewhere in the insert path) and simply select a 2 dB boost! Some of its other applications can also be great on stereo signals – switching the left- and right-hand sides, for example, or mono'ing a stereo mix (try leaving an instance across the stereo bus) to check its mono compatibility. Also, if you need to phase invert any microphones (a snare bottom, for example), simply activate the Phase Invert option.

The Multimeter plug-in is a great addition to any stereo bus, providing up-to-date information on the track's spectral properties (using a 1/3 Octave Spectrum Analyzer), alongside phase characteristics courtesy of a distinctive Goniometer. Use the spectrum analyser to get a better grasp on the spread and range of frequencies in your mix. If you're using monitors with a limited bass response, the lower spectrum information (100 Hz

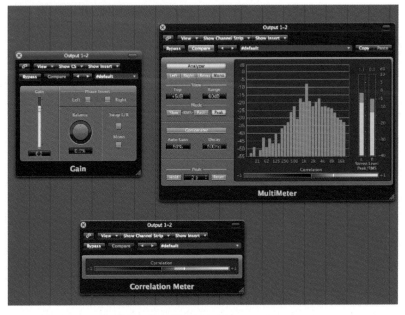

Many of Logic smaller plug-ins can aid the effective workflow of a mix – from the Gain plug-in, to the range of metering options.

and below) can be crucial to keeping an eye on any excessive activity. The Goniometer provides some indication of the stereophonic properties of your mix – a strong mono image is indicated by a clear line up the middle of display, while excursions into stereo are indicated by movement to the left and right.

The Correlation Meter – available as a separate plug-in or, on either of the Goniometer or Analyzer's screens – indicated in phase problems with the mix. If the phase is good, you should see consistent movement one side of the line (usually in the +1 region). However, if the phase is bad (indicating problematic mono compatibility) the correlation meter will move between both −1 and +1 in a continuous fashion.

8.5 Beginning a mix

With the mechanics of understanding how Logic's mixer works under your belt, let's start to have a look at the creative and technical processes of putting the mix together. However, before you start instantiating various plug-ins, take time to establish a plan for what you want to achieve. Think about how you intend to distribute instruments across the soundstage – both in respect to left to right placement (pan) and front to rear (with the use of reverb and maybe EQ). Establish what you feel should be the "lead" instrumentation – in other words, the three or four instruments that really hold the track together – alongside the other elements that should only provide a "supportive" role. Finally, identify instruments that are possibly fighting against one another – sounds that may blur the definition and energy of the track, or parts that may get lost amongst other instrumentation. Ultimately, if you're clear about some of these issues from the start, you'll have a much better chance of producing a coherent end result.

Basic part levelling

An essential part of the mix revolves around good levelling practice – put simply, balancing parts without the addition of any additional processing. Although the precise technique varies from engineer to engineer, the intention is to get a good "working balance" of sounds, some with more dominant roles than others, but all without overloading the main outputs from Logic. For example, it's easy to start putting a mix together, pushing channels into the +3 to +6 dB region only to discover that the main output quickly starts to distort. Instead, try to keep an eye on the main output level (usually on the output 1–2 channel, which should sit side-by-side the instrument channel/audio channel in the inspector), and bring in two or three of the key instruments leaving 6–12 dB or so of headroom. Unless an instrument has been recorded particularly "hot," this will probably be in the region of −4 to 0 dB.

Alongside the basic of part levelling, you might also want to consider panning the various channels to create a realistic soundstage. Think about the instrumentation arranged across the stage, creating a natural, well-distributed configuration of the instrumentation, without relying too much on the extremes of left and right (−64 and +63, respectively), and leaving a clearly defined space in the centre of the mix for important lead instrumentation like vocals (in other words, only a few channels should be left in the 0 position). Remember that as you apply effects like reverb and delay (which we'll see more of later), you can still pan these signals just like any other channel in the mix.

Applying equalization and compression

Two of the most important signal processors in Logic – and the cornerstone of any professional mix – have to be the equalization and compression plug-ins. One of main applications of compression and equalization is to provide better separation between sounds – fixing different elements into specific parts of the frequency range of a track, for example, or locking a sound better into the mix's dynamic properties. One common technique with EQ is to reduce clashing frequencies between sounds, allowing each sound to sit in its own sonic space. For example, a 200 Hz reduction on a low guitar part would facilitate a little more space for a bass, whereas the intelligibility of the vocal could be improved by a cut at 1–3 kHz on any competing instruments. Compression, on the other hand, helps iron out any dynamic inconsistencies – for example, rather than a vocal dipping in and out of the mix, it can remain consistently loud.

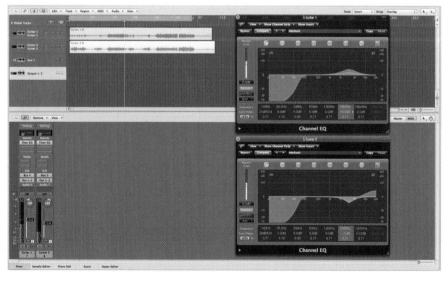

Try using equalization to create better separation between instruments. In this example, a 3 dB boost at 3.5 kHz in one guitar has been balanced out by a 3 dB cut at 3.5 kHz on another.

Plug-in boxout 4

Compressor

Along with EQ, compression forms one of the fundamental tools of a mix-down, and essentially works as a form of automated amplitude control – the compressor reacts to loud peaks in the input, and then attenuates the level of the signal accordingly. The two key parameters in this process are Threshold – rated in decibels, and governing the point at which compression begins – and Ratio, which sets the hardness of compression applied. For example, with a low threshold and ratio (-20 dB and 2:1, respectively), a slight compression would be applied, even on relatively quiet parts of the signal. Alternatively, a higher threshold and ratio (-5 dB and 10:1, respectively) would produce a stronger compression, but this would only be applied on the rare occasion the signal level exceeded -5 dB.

In addition to the key compressor settings (threshold and ratio), Logic's compressor plug-in contains a number of other parameters to further refine the effect. Attack and Release, for example, are important tools for defining how the compressor moves in and out of gain reduction, as the signal exceeds the threshold. With a quick Attack and Release setting, the compressor is set in a "fast-acting" mode – the response to loud transients (like loud snare hits) is quick and efficient, and the compressor is quick to return to its normal state once the input reduces below the threshold. As great as this sounds, however, these fast movements in and out of gain reduction can have a negative effect on sound being processed, as the compressor creates a distracting "pumping" sound. Setting a slower Attack and Release might allow the odd loud peak to slip through the net, but you'll end up with a more musical, empathetic response from your compressor.

Logic's compressor is a versatile gain reduction tool, capable of controlling the precise amplitude and dynamic range of a signal in the mix.

Peak and RMS (root mean square) modes govern how the compressor "listens" to its input. Essentially, the Peak mode responds exactly to the true level of the input, while RMS responds to an averaged level, closer to how the ear perceives loudness. With a Peak detection setting, therefore, you'll find the compressor reacting more than on the corresponding RMS setting; although its response might be considered slightly less musical. The Knee of the compressor is a useful means of creating a more graduated transition on harder ratio settings – the ratio is slightly softer (2:1, for example) ahead of the threshold, only reaching its full strength (6:1) a few decibels after the threshold.

As the overall effect of compression is to reduce the dynamic range of your input (in other words, making the loud bits quieter), you'll probably find your corresponding output quieter than without compression. Increasing the output Gain (at the right of the interface) will restore levels lost through compression. You can also use the AutoGain option of Logic to apply this automatically; although in some cases this can produce distortion.

Compression and EQ both can be applied via the insert path of a selected audio channel strips, instrument channel strips, or aux channel strips; although the EQ can also be quickly activated by double-clicking in the small EQ box at the top of the channel strip. Double-clicking the selected plug-in will then open a floating dialogue, allowing you to adjust the various parameters contained within the plug-in. Of course, bypassing effects is essential to establishing exactly what you have or haven't achieved. This can be done using the dedicated bypass control in the top left-hand corner of the plug-in's interface, or alternatively, directly from the mixer itself by option-clicking on the appropriate plug-in slot.

Adding in further plug-ins

Further plug-ins can also be inserted as part of the path of plug-ins you create on the channel strip. Remember, though, that the order in which you insert the plug-ins can have a big effect on the overall treatment produced. Even something as simple as EQ and compression, for example, can have a subtly different output based on which plug-in is placed first in the chain. More extreme effects, say for example distortion followed by EQ or EQ followed by distortion, can have completely different result given their relative position. If you need to change this order you can use the hand tool on the mixer to re-order the inserts in anyway you see fit.

One common mistake at this point is to apply too many plug-ins, possibly in a desperate attempt to improve apparent deficiencies in the source recording; "at least if it's covered in distortion and delay, nobody will notice that it's out of tune!" Try to have a clear strategy in your application of effects – not all

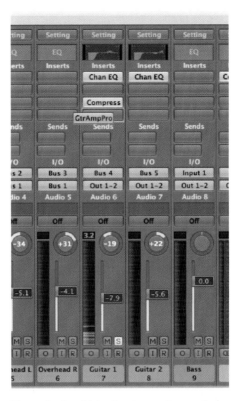

The order in which plug-ins are inserted can have a big effect on the end output. Try using the hand tool to experiment with the order of plug-in the channel strip.

tracks necessitate the use of plug-ins, and, where a more complicated series of plug-ins are used, try to make this a unique and identifiable "special" feature, rather than the norm.

8.6 Adding send effects

As tempting as it is to patch everything across a channel's insert points, it's really worth making a clear distinction between the use of true insert-based effects and send effects. The use of an insert effect implies there's no need to balance an unprocessed and processed version of the signal. EQ, therefore, is a perfect example of insert effect as there is little or no need to hear both an un-equalized and an equalized version of the sound at the same time. The same could also be said for compression. Reverb, on the other hand, is a clear candidate for being applied as a send effect, with the need to balance the respective levels of wet (in other words, reverberated) signal with the dry, unprocessed version. In theory, the greater the reverb, the more the channel appears to move towards the back of the mix.

Plug-in boxout 5

Channel EQ

The Channel EQ is easily one of the most important plug-ins in any mix, controlling the precise spectral qualities of signals passing through it. Although the Channel EQ takes a low drain on available CPU resources, it is both a powerful and a flexible tool, split into eight frequency bands with a cut/boost and frequency parameter for each band. As a good starting point to understanding your input source, try activating the Analyzer feature of the EQ. The Analyzer provides a real-time Fast Fourier Analysis (FFT) of your signal, indicating the distribution of energy across the sound spectrum – a bass, for example, should produce large "humps" formed by its fundamental at 100 Hz, alongside additional harmonics further up the harmonic spectrum.

The two extreme bands – at the far right- and left-hand side of the interface – govern the controls for the high-pass and low-pass filters within the EQ. Effectively, these completely remove frequencies above or below the given cut-off point – try using the high-pass filter, for example, to tame any excessive low-frequency energy. Moving inwards, the next two bands correspond to the shelving EQs, which are somewhat comparable to the treble and bass controls on a conventional hi-fi. Use the shelving EQ for general sweetening activities, creating the familiar "smiling" EQ curve with a boost at around 80 Hz and 12 kHz, respectively.

The remaining bands work as traditional parametric EQs, with a cut/boost control, frequency setting, and a fully variable Q parameter. Q sets the width of cut and boost, and therefore the resultant amount of cut and boost to harmonics near to the EQ's selected frequency. A wide Q is a useful way of shaping more "general" qualities of the sound – maybe a lack of bite in the upper mids, for example, or an overall woolliness in the low mids. Use a tighter Q where your frequency issues are more specific – like a boomy resonance on an acoustic guitar, for example, or prominent problem harmonics.

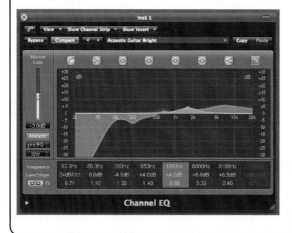

Channel EQ is one of Logic's primary mixing tools: with four fully parametric bands, two shelving controls and two filter sections.

The first step to creating a send from any given channel is to double-click in one of the available send slots on the strip and select from one of the 64 available bus destinations. Raising the level on the small send pot will then bleed an amount of the channel's signal via the bus to the appropriate aux channel strip. The aux channel strip should then have the corresponding plug-in (an instance of Space Designer, for example) patched across its insert path, so that signals entering the aux channel strip are effected accordingly. With the mix parameter on the plug-in set to 100% wet, you can now control the level of reverb fed back into the mix, using the aux channel strip's fader, or alternatively, using further plug-ins across the aux channel strip to further process the sound of the reverb.

A basic configuration of send effects – the channel bleeds of an amount of its post-fade signal, via a bus, to an aux channel strip. The aux channel strip is configured with a reverb on its 100% wet setting.

Both reverb and delay are important tools in defining the spatiality of your mix – in other words, the front-to-back perspective of the soundstage. Having worked carefully on your level and panning, you should have initially formed a good representation of your left to right soundstage. However, with the application of delays and reverb, you finally have a chance to set sounds forward and backwards in the mix – a dry vocal, for example, sat squarely at the front, or a distant keyboard pad drifting towards the back of the soundstage. One really useful technique is to place two to three contrasting instances of Space Designer to establish a front, mid and rear acoustic to in your mix. Use some of the shorter reverbs on close sounds, or rhythmic elements (like drums or rhythm guitar), that don't suit long reverb settings. The longer settings, on the other hand, could be reserved for a few unique sounds that really benefit for longer reverb tail, and a more distant mix placement.

One important concept of the sends is whether they are configured for either pre- or post-fade operation. In a mix, the most common example (and the one Logic defaults to) is to use the sends in post-fade mode – that is, the send happens after any adjustment in level for the particular channel. If, for example, a fader level is brought down in the mix, its corresponding amount of reverb is also attenuated – in effect, preserving the ratio between the two sounds (dry

Use two or three complementary reverbs to help define the front-to-back perspective in your mix. Additional delays can also help define the "spatial" dimension.

and reverberated). Pre-fade sends (as we saw in Chapter 4) are usually associated in the creation of headphone or cue mix. To adjust between pre- and post-fade operations simply click and hold on the send and adjust accordingly.

8.7 Combined processing using aux channels

Having seen one application for aux channel strips – that of applying send effects – let's have a look at the other ways in which they can be applied during a mixdown. By changing the output option of any audio or instrument channel in the mix you can also decide to route it to an aux channel strip ahead of the signal reaching any of the main, physical outputs from Logic. Used in this way, you can combine a number of signals – say, the various mics positioned around a drum kit, for example – to a single fader, with the option to control both its level and the application of additional processing en-masse.

The use of so-called 'bus processing' across aux channel strips can be a great way of locking together the principal components of a mix, as well as helping groups of sounds gel in an effective way. Many American rock engineers get a great deal of mileage from this technique, combining compression on a channel-by-channel basis, alongside compression across a collection of aux channel strips. The result is a mix that "pumps" in an empathetic way to the source material, providing a real intensity and sense of loudness brought about through the reduction of dynamic range.

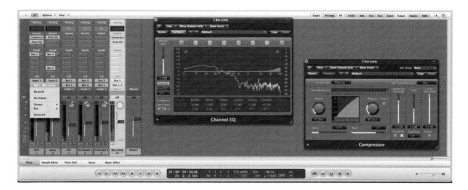

Bus processing can be achieved by sending a number of channels (via a bus) to an aux channel. On the aux channel, insert the required equalization or compression, for example, so as to process the "group" of sounds.

However, if you do intend to apply a lot of compression on your mix, it's well worth making sure that you use a variety of different compression techniques as a means of forging the identity between different sounds (rather than the track just sounding squashed) as well as matching the style of compression to the instrumentation you're processing. Try using low ratios (1.5:1 to 2:1) and low thresholds as a means of massaging the sounds into place, whereas harder ratios and higher thresholds (6:1, −5 dB) could be a great way of simply slicing a few loud transients of a drum sub-mix. With any bus compression, you'll also need to pay close attention to the Attack and Release times of the compressor, with slower settings (especially on the release) providing a more musical result.

Plug-in boxout 6

Convolution Reverb and Space Designer

Convolution has quickly become the accepted standard in producing professional-grade reverb treatments. The technique works by taking an acoustic sample, known as an impulse response or IR, from a given room or indeed, a hardware reverb unit, in question. The process of convolution then simply takes the short IR file and mathematically "folds it over" the source signal, effectively recreating the effect as if the sound had been recorded in the same space, or through the same reverb processor. For that reason, Convolution reverb can sound astonishingly realistic; although this can be at the expense of your available DSP resources. Ultimately therefore, it makes sense to run just a few instance of Space Designer, patched across bus faders, than lavish multiple instances on individual channel faders.

As part of the Logic Pro install, Space Designer comes with an impressive Library of pre-recorded impulse response files (stored under Local/Library/ Application Support/Apple/Impulse Responses). You can load these simply by scrolling through the presets, or loading them in via the Load IR option to the right of the IR sample switch. Additionally, you could choose to make use of one of the growing number of third-party IR libraries, like Spirit Canyon's Spectral Relativity, or indeed, your own custom-sampled IR files.

Based on unique samples of the original spaces, Space Designer's convolution reverb engine produces some stunningly realistic reverb treatments.

Like a growing number of the professional convolution reverbs, Space Designer offers a surprising amount of control and flexibility with effects you can achieve with it. Once an IR sample is loaded, you can immediately adjust both its volume envelope and its filter characteristics, assuming the filter is activated. Besides being able to produce some subtle modifications (rolling off the high-end, for example, with a touch of low-pass filtering), these tools can be great for abstract ambience treatments – try using the reverse setting and some extreme filter movements with lots of resonance! If you need to shorten the reverb tail, adjust the length parameter. Alternatively, to create a longer, darker reverb, adjust the sample rate into one of its slower settings. This can also be a great way of producing longer reverb times, without maxing-out your DSP resources.

8.8 Using groups

As alternative to bussing sounds together via aux channel strips you can also make use of Logic's Group feature – as we've already introduced, with the concept of edit groups in Chapter 5, Audio Regions and Editing. Taking the concept further, we can also use Groups to lock together collections of channels, say a group of drums, for example, or backing vocals. But, given the aux channel strip bussing system, what use is an additional means of ganging faders?

As simple as the bussing system is, there's one major conceptual problem – that of post-fade send effects like reverb. Imagine a collection of kit sounds all being sent to an aux channel strip, with a snare also making use of a generic reverb (also shared by the vocals and strings) on send 8. If the level of the kit were then reduced on the aux channel strip fader, you'd hope the reverb level would also be reduced accordingly, but this is not the case. As the send happens before the aux channel strip, any subsequent level changes will have no effect on the reverb – effectively the snare will get "wetter" as the level of the kit falls. Of course, one simple solution to this would be to route the reverb to the same aux channel strip, but this would then have a knock-on effect on the vocal and strings reverb!

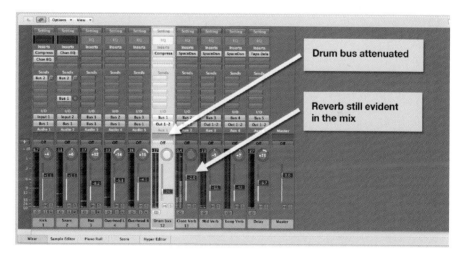

Shared reverbs can make the use of aux channels for grouping purposes problematic; although the fader level of the drums group is reduced, the reverb level stays the same.

So, here we see a perfect example of the application and benefit of grouping channels as apposed to bussing signal via aux channel strips. Grouping retains all the important gain structures of the original mix – the ratios of reverb are retained, even as the level of the group is reduced. Bussing still has its place, though, as there's no way to apply effects like compression and EQ onto the group. So, in theory, a good mix may well involve combination of both grouping and bussing via aux channel strips – grouping to "lock" collections of faders, and bussing via aux channel strips to provide grouped applications of effects!

This example uses both groups and bus processing. The Group allows us to better control the level of the drums in the mix, while the aux channel is solely used for processing rather than level control.

Looking more closely at the Group Settings dialogue, you can see a number of additional features that could potentially improve the speed and efficiency of your mix. Besides obvious grouping controls like volume and mute, you can also link the send controls (this could even be helpful in quickly modifying cue mixes as you're recording) as well as linked selection of Automation Modes. However, one vital keyboard shortcut to use in conjunction with the Groups is the Toggle Group Clutch command. By default, this is configured onto Apple+G, and can be used to temporarily disable all groups – a great way of quickly modifying an "internal" balance within any number of groups, without having to laboriously enable or disable the relevant groups.

Knowledgebase 2

Parallel Compression

Parallel compression has become a real buzzword in mixing circles, but what is it, and can Logic be realistically used to apply it? Unlike the traditional insert-based approach to applying compression, which works on the assumption that you'll only want to hear the 100% compressed signal, parallel compression offers the unique possibility of hearing both compressed and un-compressed versions. Although parallel compression is arguably less effective at reducing the dynamic range of a signal, it is an excellent way of combining both the hyped-up sound of heavy compression, with the natural dynamics and "air" of an uncompressed recording – yes, to use the cliché, it really is the best of both worlds!

One really simple way of creating parallel compression is to copy the source track onto two adjacent channels. On the one channel, keep the signal uncompressed and open, while on the other, try setting up really juicy, pumping compression. Don't worry about the integrity of the signal – pick a suitably tough ratio (6:1 or harder!), relatively quick Attack and Release, and let gain reduction pump well into the 6–10 dB range. Some engineers will even EQ this channel, picking out the extreme highs and lows (80 Hz and 12 kHz), as well as applying a little tuck in the mids. With the compression set up, combine the two channels together, mixing in the compressed version to add "balls" and body to the uncompressed track.

Another interesting approach is to use the compressor almost like a send effect – blending any number of sounds through to the same heavy compression setting. Try doing this on a drum mix, sending mainly the close snare and kick drum through to the paralleled compression. Again, pick some suitably pumpy settings, maybe even a touch of extra ambience, and sit the bus in the mix to add the required amount of weight and importance.

In this example of parallel compression, a bus send is being used to add compression using the same routing technique as a reverb send. Use a heavy compression setting, and then bleed in the required amount of compressed signal to add "body" without compromising transient detail.

8.9 Working with channel strip settings

As you're building up the mix, you may well find that certain channels require similar settings to that of others – maybe you've compressed and equalized one backing vocal, and you want to apply the same setting to corresponding backing vocal track. Clicking on the small arrow to the right of the Insert label will allow you to open up the channel strip settings menu. Try using the copy/paste option as a simple means of duplicating the channel settings on a number of channels. Alternatively, if you have a number of favourite channels strip settings – maybe a particular compression and EQ on a given vocalist, for example – you can also save the presets off, as you would any of the other plug-in presets. These favourites could then be re-called at any point, on any given song.

Use the channel strips settings as a quick way of moving one channel's plug-in configuration to another. The presets feature also allows you to do this between songs.

8.10 Automation: the basics

Despite the considered application of compression, you may still find the demands of your mix changing from verse to chorus and vice versa. Ultimately, although one balance may work at a given point in time of the song's development, the same might not be true as further instrumentation enters and the musical qualities of the track change. In the "golden days" of recording – before computers stepped out of the office and into the studio – it wasn't uncommon for a mix to involve "all-hands-on-deck": the engineer, assistant engineer, and even the members of the band, all frantically pushing faders up and down to shape the mix to the dynamic of the song.

Logic's Automation provides complete control over qualities of the mix throughout the duration of the song – including basics like volume, pan and mute, and more advanced options like plug-in parameter automation. The golden rule to remember with automation is only to apply it toward the ends of a mix – once you've established the principal balance, equalization, compression and effects usage, only then should you begin to turn to automation. The problem is that any mix changes after automation, although not impossible, can be quite a headache to perform – potentially requiring you to re-write or delete existing automation moves.

8.11 Track-based versus region-based

One potentially confusing aspect of automation is that Logic contains two contrasting methods for applying automation – track-based automation and region-based automation. In truth, region-based automation – where automation data is contained within the audio or MIDI region itself – is actually a throwback to the earlier versions of Logic, and has largely been retained for the purpose of backwards compatibility, although, to be fair, it still has its uses. Track-based automation – where automation data is recorded on a separate multiple tracklanes to that of the audio region – offers a far more flexible means of creating and editing automation data as well has having the distinct operational benefit of not being tied to the regions in questions. Editing or moving a guitar solo, for example, won't necessarily disrupt the automation data that accompanies it.

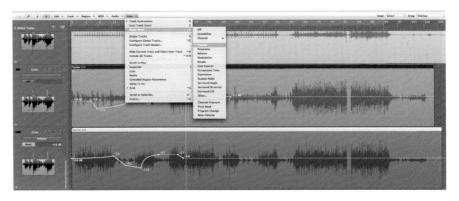

Two different approaches to automation – the old-fashioned Hyper draw (top) and the standard track-based automation (bottom).

Thankfully, Logic does provide a means of transporting one automation type (region-based or track-based) to another – making the combination of the two approaches quite a powerful feature. For example, you could write the automation using the superior track-based controls, and then convert this information into automation data that is stored with the region. For now though, we'll take a look at how track-based automation works, and then consider the region-based case later on.

Plug-in boxout 7

PlatinumVerb

PlatinumVerb is a valuable alternative to Space Designer, either for Logic Express users (who don't have access to Space Designer) or for situations where you're running slightly low on available DSP resources. In contrast to Space Designer, PlatinumVerb places a surprisingly low drain on the CPU; although correspondingly, you may find its output slightly less authentic than Space Designer. Used carefully however – on short settings with drums, for example, or any instrumentation sitting lower down in the mix – it can produce surprisingly effective results.

One of the best things about "modelled" reverb, as apposed to convolution reverb, is that it offers precise control over the sound produced in the virtual room. In the case of PlatinumVerb, it divides the reverb up into two key stages – the early reflections; as sound initially bounces back from the

If you're short of processing resources, or don't have access to Space Designer, PlatinumVerb makes an excellent alternative source of reverb.

walls, and the more diffuse reverb tail as the reflections merge together to create a definable "trailing-off" to the sound. Try moving the Balance slider over each side to fully understand and audition the effects obtained in each stage.

Using a predominant mix of early reflections can be a great way of thickening up drum sounds by producing a noticeable, distracting reverb tails. Experimenting with the room shape, stereo base and room size will change the qualities of reflections produced, either by creating a tight centralized reverb or by a wider, more expansive set of reflections. When it comes to the reverb tail, try exploring the density, diffusion and reverb time parameters. On the whole, most conventional reverb effects tend to stay within the realms of 1–2 s, and, not surprisingly, this is where PlatinumVerb seems to sound at its best. Density and diffusion, respectively, govern the spacing and randomness of the taps. Try using both in their lower setting for an effect like Spring Reverb (great on electric guitars), or use a higher setting for a smoother sounding reverb.

8.12 Automation modes

To engage a channel into automation you'll need to change automation mode according to the way in which you intend to write automation data into Logic. By default, all faders are set to their Off position, just above the pan pot – this means that any existing automation data are ignored and the faders can be freely moved or repositioned (without fear of Logic snapping them back!) at any point. Engaging any of your faders into an automation write mode (write, touch or latch) will allow you to begin writing data based on the current song position. Interestingly though, the "recording" of automation data is independent of the transport's record switch; in other words, you only need to be in play mode for automation to be written.

First of all, let's take a look at the various modes used to record automation data.

Write

Think of this mode as the most dangerous! Any fader engaged into write mode will record data onto the automation lanes, even if there's existing data, so this should really be used with care. Write, however, can be a valid way of deleting and replacing automation data in one pass.

Touch

Touch is the safer way of writing automation data, as it will only engage into writing data once the fader has been "touched." Release the fader – even as the track is still playing – and the fader will return the previous recorded position, and carry on reading any existing automation data. Touch, therefore, can be great in creating a few strategic nips and tucks in your mix – either briefly lifting a phrase out for a few seconds, or pulling back any part that dominates the mix.

Latch

Latch mode is comparable to touch, in that a fader is only engaged into writing data once it has been touched. However, when the fader is released, it will continue to write data – potentially erasing any existing moves – and the level it was last left at. Latch can be a useful mode when you need to raise or lower a level of a part and then leave it at that level for the remainder of the song without having to constantly "hold" the fader in place. Stopping the transport will, of course, stop the writing of automation data.

Read

Read is a "safe" automation mode, where any automation data is read back but no further moves can be written – indeed, if the fader is moved for any reason it will promptly snap back into place! On the whole, most users will tend to leave faders in touch mode as they're automating, allowing them quickly engage the writing of automation data without having to constantly switch modes. However, to reduce the risk of overwriting certain aspects of the mix, it's best to switch all faders back to read mode once you have finished automation.

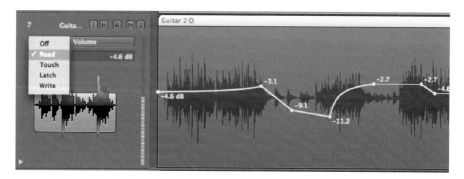

Each of four different automation modes used in Logic have their own impact on how you read and write automation data.

Although the description of different modes has assumed you're working with the channels faders, it's important to remember that once a channel has been placed into write, touch, or latch, any movement in its accompanying parameters – including mute, pan or any of the plug-in parameters – will also be recorded. This allows any part of the mix to be automated with the same degree of flexibility as the channel faders – for example, the reverb time of a vocal reverb could be swelled going into a "larger" chorus, or the feedback of a tape delay unit could be modulated to create some dub-inspired delay treatments on the end of a line.

Plug-in boxout 8

Using Delay

Although most users tend to perceive reverb as the main tool for defining spatial qualities of a mix, it's also surprising to realize just how much spatial interest can be added with something as simple as a delay line. Logic comes with three different delay plug-ins – sample delay, tape delay and stereo delay – facilitating a range of different delay treatments.

As the name suggests, tape delay is modelled on the type of delay effect produced by classic tape-based delay effects like Roland's Space Echo or the WEM CopyCat delay. As these processors use tape to produce their delays, the effect has a characteristic "dark and dirty" sound to it. By default, the plug-in works with tempo divisions – simply select the required division (semibreve, crotchet, quaver, and so on) and the delay will appear in time with your track, even if the song's tempo changes. Use the Groove slider in the extreme settings (33 and 75%) to change the division to dotted note values. You can also achieve some great slap-back delays (a effect famously used on early rock'n'roll vocals) by using a semiquaver setting and sliding the Groove parameter down into the region of 33–50%.

Feedback sends a proportion of the sound back on itself, effectively creating a regenerative delay effect. On the tape delay, feedback settings in the region of 50–100% will appear to "hold" the delay, with an increasing amount of distortion and grit on each repeat. Stereo delay works on the same principle as tape delay, although it deliberately avoids the tape-based colouration, in preference of a cleaner digital repeat. You can also set different delay time for the left- and right-hand side of the stereo image.

Logic's simple delay plug-ins can be a surprisingly effective tool for mixing – from rhythmic delays effects to subtle forms of "slap-back" ambience.

In contrast to the other delay plug-ins, sample delay only deals with incredibly small delays times, measured in samples. As a rough guide, 44 samples equate to 1 ms of delay, so even at its maximum setting (4000 samples), sample delay only gives us about 90 ms to play with! So why use this plug-in? Well, sample delay can be beneficial in applications fixing minute time delays caused when mics are widely space apart – applying sample delay to correct these anomalies and create a more phase-coherent image. If you're less technically inclined, try using a small amount of sample delay on some drum room mics to simulate the effect of sound reflecting further away from the kit.

8.13 Viewing and editing automation

By default, Logic hides the display of automation data – otherwise an arrangement could soon become cluttered. However, once you've started recording a few moves, you might want to see how the mix is beginning to shape up. Select View > Track Automation to display the current recorded automation – you might also want to toggle this using a keyboard shortcut (the default is A), so that you can quickly switch automation viewing on and off.

You'll need to enable the automation view mode to see or edit the moves you've recorded.

With the automation view engaged, you'll notice some important changes to both the tracks and the arrangement itself. Looking first at the tracks, you should now see the automation mode indicated on them (read, touch and so on) alongside the current viewable automation parameter – clicking on this should allow you to scroll through all the parameters available for automation. You'll also see a small bar graph meter indicating the current fader position. Note that this can be freely modified and controlled, just like the "real" fader as part of the mixer, and can be a great way of adding in a few cunning automation moves.

Where multiple plug-in parameters have been automated, you can also add further track lanes into the equation, simply by clicking the small arrow towards the bottom right-hand corner of the track name in the track list. Alternatively, option-clicking will open as many automation lanes as is required to display all the automation data currently recorded.

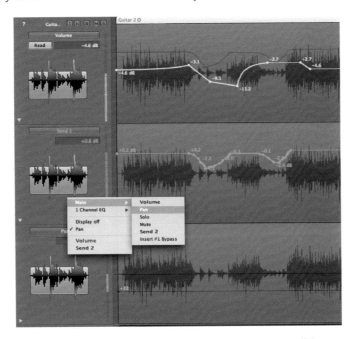

Press the small arrow on each track lane to open up a parallel automation line, making it clear to see the moves on several different parameters on the same track.

Automation data itself is displayed as a series of nodes, which can be manipulated using the usual pen, eraser and arrow tools. As you'd expect, the pen tool allows to draw new automation moves, while the arrow can modify existing node points, or by double-clicking add new nodes into the equation. Indeed, in many situations, it may be quicker and easier to draw in a couple of automation nodes than to laboriously write in a series of moves. Manually placing the nodes can also be a good way of setting in long automation events – a long filter sweep over 30 bars, for example – or when events, like fade-outs, need to happen at a precise time.

267

Plug-in boxout 9

Vocal Processing

Vocals can be one of the trickiest parts of a mix to get right, so it's great to see a range of plug-ins suitable for vocal processing in Logic. Away from the core plug-ins that we've already covered – including EQ, compression and reverb – you'll also find the DeEsser and Pitch Correction plug-ins helpful in crafting that polished vocal performance your mix requires.

Logic's DeEsser is a handy way of taming any excessive sibilance (like the letters s and t) that might have been accentuated by a poor choice of microphone. The DeEsser works by analysing the input and then applying a selective amount of gain reduction whenever problematic sibilance is heard. The Detector part of the interface is what should be used to spectrally locate the sibilance in the vocal. Set the Monitor mode to Det (detection) to tune into the specific frequency your signer's sibilance is occurring (usually somewhere in the region of 6 kHz). Now, flick to the Sens (sensitivity) monitor setting and adjust the sensitivity so that it only flicks on when the sibilance occurs. Moving the monitor to its OFF position, adjust the Suppressor parameter to the same parameter to that of the Detector and increase the strength to get the required amount of sibilance reduction.

The Pitch Correction plug-in is loosely based on the infamous Antares AutoTune plug-in, and provides an on-the-fly means of correcting intonation problems in a vocal – or indeed, any other monophonic performance. At extremes, using the response setting on its 122 ms setting, the plug-in can be forced to produce the clichéd, quantized vocal effect so carelessly abused in the late 1990s! On softer sensitivity settings however, it can be a useful way of taming any problematic pitch drifts. To achieve the best results with the plug-in you'll need to specify the key and scale the song is in. A number of presets for this are available, but you can also manually switch the notes in and out by clicking on the accompanying keyboard.

Logic's DeEsser and Pitch Correction plug-ins can aid a number of problems in relation to your vocals.

When two nodes are placed manually, you can also make use of unique automation curve tool and as a means of adding a degree or "curvature" to the line. Simply click and hold on the line between the nodes, using the automation curve tool, and drag above or below the line, or from side-to-side, to create one of four different adjustable curve shapes.

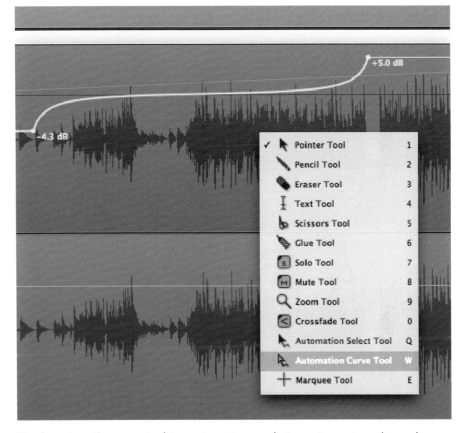

Use the automation curve tool to create neat curves between two automation nodes.

Where ranges of moves need to be moved or duplicated, you can also drag-enclose a number of nodes simply by holding Shift as you rubber band a group of nodes. With the nodes selected they can then be moved en masse, or with the option key held down, duplicated to a new position. One other potential lifesaver, in case you've written in the automation moves but feel the need for them to be a couple of decibels higher or lower, is Logic's scaling feature. To scale a track's automation moves use the Apple key and drag up or down on the small bar graph meter as part of the track list. Look carefully at the selected values, and you should be able to spot the corresponding changes you've made.

Knowledgebase 3

Saving CPU: Freezing and More ...

At some point in the mixing process it's highly likely that you'll start to hit dreaded inevitability of a CPU overload. Dealing effectively with this problem could make the difference between a half-baked mix and a distinctive mix, using appropriate strategies to save CPU resources without compromising on sound quality. First off, it helps if you're clear as to which plug-ins are the real CPU monsters in your session. For example, synthesizers like the ES2 and Sculpture can take a big quota of resources, while the EXS24 (assuming it not making lots of use of its filter) is surprisingly processor efficient. In respect to plug-ins, Space Designer can be really CPU-hungry, especially if you're using longer IR files in excess of two seconds, while the Channel EQ is almost negligible on today's Intel-powered machines.

Arguably the first technique, therefore, is to look for any optimizations you can apply to the session that might improve overall CPU efficiency. For example, try turning off unused oscillators in the ES2, or removing excessive Space Designer use on inserts, in preference for a few instances on aux sends.

The next step is to use the track freeze option. Freezing effectively creates an audio bounce of the track in question, completes all its inherent plug-in settings, and then deactivates the plug-ins accordingly. If you then try to edit any part of the track – either opening an instrument or plug-in or repositioning a region – Logic will remind you of its "frozen" status. If you do need to carry out the edits, simply unfreeze the track, make the modifications, and then refreeze.

Freezing itself is carried out with the small freeze icon as part of the track header. If you can't see the icon, select View > Configure Track Header and then add the freeze button into the set of options. Note that the freeze will occur across the full duration of the project, so it's worth moving back the project end marker to the real finish point of your track.

As an alternative, you can also carry out an audio bounce yourself, simply rendering complex virtual instruments as audio regions, saving off the instrument settings, and packing away the MIDI data (just in case you need to go back). This can be an effective solution earlier in the project where you intend to do lots of structural re-arrangement with the regions at a later point.

Use the freeze function to render an off-line version of the track and so release valuable CPU resources.

8.14 The automation menu options

Besides the various graphic tools for editing automation data, there's also an accompanying automation menu (Options > Track Automation) containing various options to delete moves, as well as some intriguing option to move data back and forth between track-based automation and region-based automation (as previously discussed). Although it is, of course, easy enough to delete selective parts of automation using the eraser tool, the menu options make quick-and-easy to create "blank-slate" on individual automation lines, tracks, or indeed the whole song itself. Certainly, in situations where you might inherit a previous song file for a new project, this can be a great way of clearing out problematic automation moves.

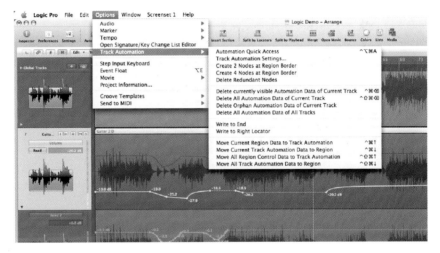

The track automation menu includes several powerful features, including the ability to move automation between its native track-based form and the region-based solution of Hyper draw.

Moving track-based automation to region-based automation is useful in situations where you'd like the automation permanently attached to a part, without having to continually specify for the automation data to be moved every time you re-align or duplicate the part. Possibly the best "real world" application of this would have to be the use of automation to control filter movements – maybe you've written a couple of distinctive filter movements that are as much part of the musicality in the region as the notes contained within the MIDI sequence. By switching the automation over to region-based (Options > Track Automation > Move Current Track Automation Data To Region), you effectively lock the movements into the part, allowing yourself to forget about whether the data are moved or not. Of course, at any point, this same data can be brought back to track-based automation using Options > Track Automation > Move Current Region Data to Track Automation.

Knowledgebase 4

Compression Circuit Types

The qualities of the different compressors are something that tend to excite many professional engineers, mainly as the different approaches to circuit design can achieve some radical differences in the types of compression achieved. Ultimately, this means that certain compressors tend to suit themselves to specific applications, bringing plenty of "character," as well as gain control, to the input they're processing. In the example of Logic's compressor plug-in, therefore, we're provided with Platinum, ClassA_R, ClassA_U, VCA, FET and Opto.

Although the exact models from which these circuit types are derived aren't supplied, it is worth noting some of the key differences between these different options and how best to apply them. Opto represents the oldest compressor design, based on early models that used optical cells as part of their gain control circuitry. This unique design resulted in a degree of latency – both with respect to the Attack and Release on the compressor – that tends to deliver a more "musical" compression, as apposed to a harder, more aggressive gain control. The Opto circuit type, therefore, works well with bass sounds (that don't suit fast Attack and Release times), vocals, or anything that you want to retain a degree of musicality and lightness with.

The FET design, on the other hand, used Field Effect Transistors to create a more heavy-handed compression, particularly good at catching transients. The FET circuit model therefore, and to our ears the ClassA_U, can produce some really effective results on drums, especially when used across overheads. In this application, don't be afraid to use low threshold setting with "pumping" Attack and Release times for an aggressive, almost low-fi compression sound.

Although this provides a theoretical background to select the circuit type, the best approach is to use your ears. Try configuring some basic compression settings and then flick between the different compressor models to hear the marked differences in how they sound.

The different circuit types approximate the unique sonic behaviour of many classic types of compressor, like the Urei 1176 or LA-2A.

Walkthrough

Adding Compression

Step 1:

Insert a compressor across the instrument you want to process and start by establishing some basic settings. Working from the default positions, try finding a ratio and threshold setting that works for the instrument you're trying to process. For example, for a gentle compression use a medium threshold with soft ratio (1.5:1 through to 2:1) yielding about 2–3 dB of gain reduction (another word for compression). For a harder compression effect, consider bringing up the ratio (4:1 or more) and lowering the threshold to achieve 6 dB or more of gain reduction.

Step 2:

With these basic settings established, you can now start to refine the compression a little. Try adapting the Attack and Release settings to best suit the style of compression you want and the sound you're trying to squash. Slower settings (Attack 40 ms, Release 400 ms) tend to create a more natural progression in and out of gain reduction; although you might find the occasional loud transient slipping through the net. Faster Attack and Release settings (Attack 0–10 ms, Release 100 ms) produce a "pumping" effect, which tends to work well where you want the compression to sound more noticeable.

Step 3:

As compression leads to an overall loss in level, you need to raise the output to restore the overall signal to its original peak level (although, of course, the signal will be more compressed). Try bypassing the compressor, noting the meter readings, and then using the Gain parameter (with the compressor active again) to restore the original level. Listen carefully to the compressed sound in the mix – is there enough compression to sit the instrument correctly? Is the compression too obvious? Further fine-tuning (maybe increasing the ratio or softening the Attack and Release) will optimize the compression for the instrument's position in the mix.

Logic tips

Plug-In Delay Compensation (PDC)

Any plug-in added into a channels signal path will create a small amount of delay, or latency, through the extra processing required in producing the effect. Thankfully however, Logic includes a feature to compensate for any delays incurred through plug-in processing, called PDC, available under the general tab of the audio preferences (Preferences > Audio). For users using standard audio unit plug-ins (either Logic's own or from other developers), stick to the "audio tracks and instruments setting." However, if you are running a processing-accelerator system like Universal Audio's UAD-1 or TC Electronics' PowerCore system, you may notice delays building up when you start to use buses to apply UAD-1 or PowerCore plug-ins. In these situations, change the preference to its All setting – Logic should now play the entirety of the session in time.

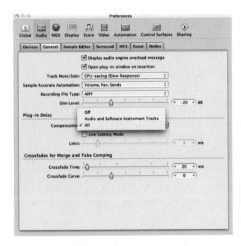

Changing the plug-in delay compensation settings is important for users of processing-accelerator system like Universal Audio's UAD-1 or TC Electronics' PowerCore system.

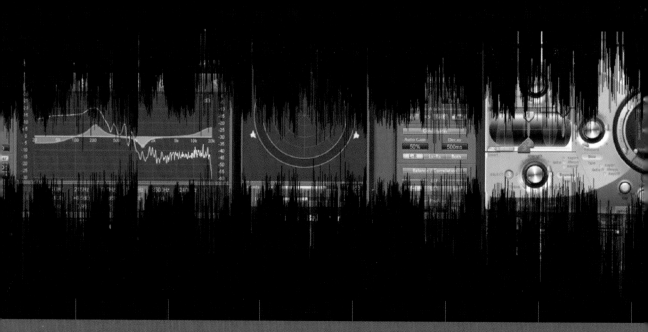

Mastering in Logic

9.1 Introduction

If you've ever tried burning a few tracks onto a CD, you'll be well aware of the challenges of producing a CD that sounds comparable to a commercial release. Even with a complete mastery of the production process, and some great-sounding final mixes, your CD could still sound weak and comparatively amateurish. Although OSX and iTunes offer integral CD creation and burning, the fact is that professional musicians and bands will invest a significant amount of money and experience in turning their finished mixes into a final product. So does this mean that users of Logic can't enjoy the same degree of finesse and polish? Well, with a little know-how, and the audio tools of Logic, you too can produce a release-quality CD master.

The term Mastering describes the process of compiling and editing several (possibly contrasting) recordings, applying some form of audio "sweetening," and assembling these to produce a final Red Book "production master" CD. Traditionally, mastering has necessitated the use of a separate facility (other than the music studio), specially equipped with mastering equipment including multiband compressors and high-end mastering EQ, as well as dedicated workstations like Sonic Solutions. Nowadays, however, the world of media is a lot more demanding, so it's not uncommon for musicians, bands and composers to master themselves. The bar, it appears, has been raised – but Logic is certainly up for the challenge!

9.2 Different approaches to mastering

Mastering itself can be divided into several objectives. First, of course, is the sweetening we most commonly associate with a commercial CD – in other words, the use of Compression, EQ and a host of other processes across the finished two-track master. Second, tracks will also need to be edited – setting correct start and end points, for example, or placing any fade-ins or fade-outs as required. Finally, the finished files or regions need to be ordered for the CD, with appropriate markers to define the tracks and index points that appear on the CD. With these three objectives, you can master in the Logic universe.

Technique 1: iTunes

The first technique, and the one most Logic Express users are used to, is the idea of applying mix sweetening, edits, fade-outs and dithering all options inside the main Logic application. The finished files are then rendered (using Bounce to Disk) or exported as 16-bit 44.1 kHz files ready for compilation. The compilation process, however, requires the use of other software – either iTunes (Apple's integral audio CD burning tool) or dedicated Red Book software like Jam.

Audio masters, exported from Logic, can be assembled into a finished CD using iTunes or other suitable Red Book standard software.

Technique 2: Burn from Logic

For the quickest and the most integral solution to the mastering problem, you can now burn CDs directly from Logic. Again, sweetening and edits can be applied directly to the main application. The burning of CDs, however, is carried out from the Bounce menu – rather than rendering an audio file, you'll burn the finished mix onto a CD. This method is arguably the quickest way of creating a CD, especially with just one or two tracks needing to be burnt, but it doesn't offer the most flexible solution in the long run.

For a quick CD, use Logic's integral CD Burn feature, as part of the Bounce to Disk option. Although this method is an easy way of burning single tracks (a rough mix, for example), it can be too restrictive for professional mastering.

Technique 3: WaveBurner

The best technique – although only available to users of Logic Studio – is to use Apple's dedicated Red Book application called WaveBurner. Originally developed by Emagic, WaveBurner has previously been sold as an individual programme, but is now included as a standard in Logic Studio. In essence, WaveBurner shares many features with Logic (including plug-ins like Multipressor or De-noiser) alongside tools specifically dedicated to the process of assembling a Red Book standard CD. The real advantage, however, is the way in which you can experiment with the order and sound of your CD – with automatic track crossfades, individual plug-ins for each region (or track), and the important Red Book options like ISRC codes and CD-Text.

9.3 Bounce to Disk

Whether you are mastering in Logic itself, or using another application (like WaveBurner), the first step will be to render your Logic mix as an audio file. The temptation to apply audio sweetening or fades in the main Logic project can be persuasive, but this should be avoided at all costs. Ideally, if you have at least one copy of your song un-mastered at 24-bits, then it would be more suitable to take to a professional mastering engineer (when the lucrative record deal arrives!) than a home-mastered 16-bit file. Separate high-resolution files also afford you the opportunity to approach mastering your work away from the

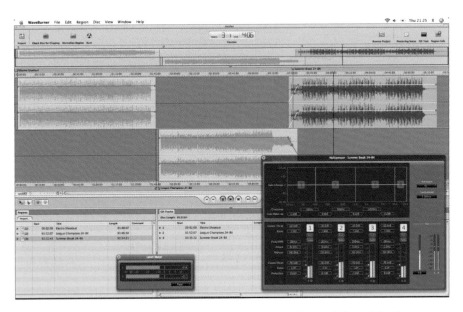

WaveBurner offers the best method for mastering and producing fully Red Book compliant CDs.

"headspace" of a mix, so that tracks sound correct in terms of the whole CD rather than on an individual basis.

The Bounce dialogue window can be accessed via the Bounce button, on the main output channel strip. Before clicking on this, you'll need to define the length of the bounce, designated by the current cycle length – without this, Logic will simply default to bouncing from the beginning of the first region to the end of the last. By defining the length manually, you can keep the best account of a few crucial factors – namely, the "hangover" at the end of the track generated by reverb tails and the allowance of a small amount of silence at the start of the audio file. At this stage it's probably best to render a slightly larger file, rather than too small, as trying to add information later (say, for example, when a reverb tail gets cut off) can be tricky, if not impossible.

Knowledgebase 1

Pre-Mastering

Officially the term "Mastering" refers to the process of cutting a master disc from which the records would be pressed in duplication. Pre-Mastering originally described the process that engineers used to prepare the audio signal for the vinyl medium. Nowadays, the term Mastering tends to be used in place of Pre-Mastering and describes the process of preparing the audio for the intended medium, ordering the music and also signal processing.

In the Bounce window you'll need to specify the file type, resolution, and dithering options. Ideally, a PCM, AIFF, 24-bit, 44.1 kHz, interleaved file is considered the best "raw data" for mastering. More important, you should check that the dithering has been set to none, as dithering is best applied at the very last stage of mastering when the word length is reduced to a 16-bit master. The Bounce itself can be carried out in Realtime (maybe you've got some live synths or effects coming into the audio mixer) or Offline – a quicker way of rendering to file by temporarily devoting all your computer's resources to the bounce process.

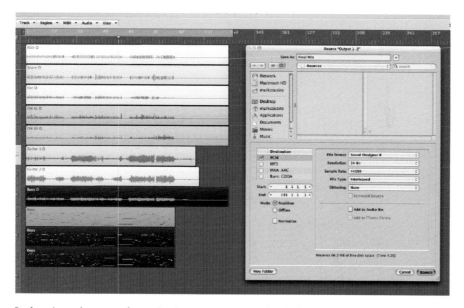

Performing a bounce of your Logic arrangement – check that the duration of the bounce is slightly longer than your track and the resolution is 24 bit.

Knowledgebase 2

Dithering

Dithering is an essential part of the analogue-to-digital conversion process. However, the process of dithering a signal also refers to the process of altering the wordlength to suit your specification. For example, if you bounce your Logic song as a 24-bit file but wish to place this on a Red Book CD, you will need to dither this down to 16 bits. The 24-bit recording will allow for a much more accurate 16-bit file due to the increased resolution, unlike the "dither" noise added in 16-bit A/D conversion. Logic employs a licenced algorithm called POW-r (Psychoacoustically Optimized Wordlength Reduction) from the POW-r Consortium LLC (http://www.mil-media.com/docs/articles/powr.shtml). The POW-r algorithm is considered one of the best dithering algorithms and is widely supported by mastering engineers around the world.

9.4 Audio mastering in Logic

With the raw data of your tracks assembled, you can begin to look at mix sweetening. Whether you're using WaveBurner or Logic, the principal plug-ins and objectives will be the same, although the exact details of their application will vary. With the various songs pulled into a master Logic session, you could place each region on a different track and experiment with various plug-ins to reach the desired sound. Ideally, the CD should present a rounded and uniform tone throughout, with a consistent "loudness" across the tracks – you could even try importing some commercial tracks as a reference to see just how far you can take things. Exactly how you achieve this will vary from track to track.

The principal tools used in mastering are well represented in Logic, with a number of dedicated mastering plug-ins, including phase-linear EQ, multi-band compression and limiting. Obviously, you can use any plug-in where appropriate, but these particular tools will be the most useful in achieving a professional sound. Plug-ins can be inserted either on the individual track's insert points (for song-specific processing) or across the main master channel strip for general application (maybe when the album needs to be limited as a whole, for example). As with any audio processing, the order of the plug-ins is vital for the end result, although the widely accepted order for mastering is EQ, followed by compression and finally limiting. Metering (applied across the main output using the Channel EQ's Spectrum Analyzer or the Multimeter plug-in) will help keep an overview of things – note, in particular, how the commercial tracks might meter differently to your own mixes.

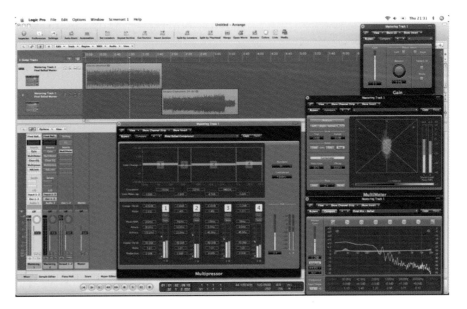

Mastering in Logic using a combination of channel inserts and the main stereo output object.

9.5 Editing fades

As well as preparing the sound of the tracks, it's also important to look at some other important mastering details, like the exact start and end points, and any desired fade-ins or fade-outs. Having already mastered audio editing in Chapter 5, you should have no problems switching over to working on whole songs rather than individual regions. When you're editing the start point, however, make sure you leave a small amount of silence and don't cut right to the beat – this allows for older CD players to de-mute, and stops the track from "jumping-out" at the start of playback. Also pay close attention to unwanted noise at either end of the track – the easiest way to address this is to use the Sample Editor's Functions menu to "Silence" unwanted noise, or use a quick Fade In and Fade Out as appropriate.

Use Logic's Sample Editor to perform basic "top and tailing" tasks, like silencing unwanted noise or creating small fades (as illustrated).

Plug-in focus 1

Linear-Phase EQ

Having looked at Logic's standard channel EQ in the mixing chapter, let's take a look at the mastering-orientated Linear-Phase EQ. Being much more CPU intensive than the Channel EQ, the Linear-Phase EQ produces a technically superior sound by removing the phase shifts (caused as frequencies are cut and boosted) that occur in a conventional EQ. Even with its CPU drain, the Linear-Phase EQ is a welcome and sonically accurate tool for

The Linear-Phase EQ removes any of the usual phase shifts that occur with conventional parametric equalization.

enhancing and shaping the timbre of a track in mastering. Operationally speaking, the Linear-Phase EQ includes the same controls as found on the Channel EQ, and the same phenomenally useful FFT analyser.

Applying EQ in Mastering requires an approach different to that of mixing. On the whole, your approach needs to be a subtle as possible – remember everything you do will be much more noticeable as you're only working with two stereo tracks! Try to keep boosts, or preferably cuts, to a maximum of ± 3 dB with a wide bandwidth (low Q parameter) – if you're going beyond this, you may well have a problem with the original mix. Overall, your track should exhibit a smooth response, with a gentle roll-off of high frequency energy. Remember though that there are other tools that can also have a timbral effect on the mix – sometimes more successful than EQ – like multi-band compression (especially on bass) or harmonic excitement (for some top-end sparkle).

Creating proper track fade-outs (i.e. fading out over the last chorus) is a little more taxing. The problem lies in the fact that adding a fade using Logic's conventional method – arrangement area fade tool – doesn't account for the use of mastering processors. As the fades are pre-effects, any corresponding change in the track's level will result in a change in compression – so as the track fades out, for example, it slowly loosens its compression. To counteract this, try performing another bounce, this time with the processing in place. The new "compressed" region can then be re-inserted into Logic (the old effect being disabled) and then the fade can be applied.

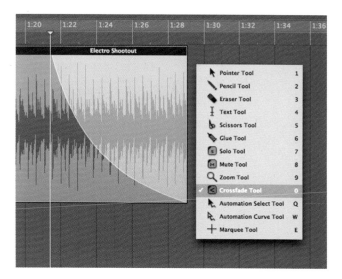

For longer fade-outs you'll need to render the file with effects, then create the fade using either Logic's region fade tool (as illustrated) or the Sample Editor's fade.

9.6 Exporting and burning

Once the tracks have been processed and edited correctly, you can begin to create the final exported 16-bit masters ready to be burnt, or indeed, you could burn the files directly from Logic. The Bounce dialogue window again comes into action – either to create the final rendered files or access to the Burn feature. Given the regions of the exact finished song length, you can use the Region menu to "Set Locators by Regions," in this way the cycle length (and therefore the Bounce) will be exactly the same length as the region. The bounce dialogue window should specify the creation of 16-bit files, which will necessitate the (final) application of Dither to smooth out any quantizing noise brought about by moving away from 24 bits. Once exported, these files can be dragged straight into iTunes (or other suitable software) ready to be burnt.

Using the "Set Locators by Regions" feature you can create a new bounce with exactly the same length as that of your edited regions.

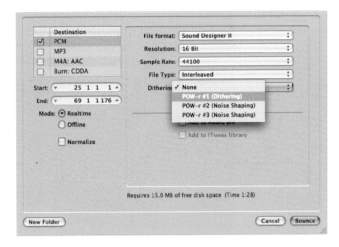

The last bounce will need to be dithered down to 16-bit resolution, ready to be burnt in another application.

With Burning now supported directly from the Bounce dialogue window, you can also burn a CD directly from Logic itself. Again, where the session contains audio files greater than 16-bit resolution, the addition of dithering is essential, which can be selected from the appropriate pop-down menu in the Bounce window. Other pertinent options include setting the write speed (keep this low to avoid any write errors that degrade the audio quality on playback) and writing the disc to a multi-session CD. When the "write as multi-session" option is selected, you'll be able to add further mixes to the CD at a later point, otherwise Logic will simply burn a single onto the CD. Although adding sessions at a later point will save the use of CD-Rs, you might experience problems with CD players recognizing songs burnt in subsequent sessions.

Quick and easy CD creation: Logic's Bounce and Burn feature.

In many ways, the "Bounce and Burn" feature is best viewed as a quick fix for creating audio CDs – especially if you intend to write more than one track to the CD at any given time. In this scenario, it might be perfectly legitimate to attempt some rough mastering in the actual project's file, rather than a separate session – simply insert a Multipressor and some EQ across the main stereo outs and away you go!

Plug-in focus 2

Multipressor and Multiband compression

Originally developed as part of WaveBurner, Multipressor is Logic's answer to the big multiband compressors (like the TC M6000 or TC Finalizer, for example) so frequently used in mastering. Unlike conventional single-band compression, a Multiband compressor splits an incoming audio signal into separate frequency bands before applying compression. Separated in this way, a complete mix is far easier to control and gel to form a finished master. In most cases the greatest amount of compression will be applied to the bottom end of the mix – keeping the bass solid and tight – with the high-end requiring a light and minimal touch.

Multipressor features up to four bands of compression and downward expansion, although to make things easier you might just want to use three bands (low, mid and high respectively). One of the most important things is to set the right crossover frequencies for the different bands, as

The Multipressor allows you to compress your master using a series of different frequency bands.

this can have a big effect on the finished compression achieved. To set the crossover point, try moving the vertical borders on the top of the interface's window – you could also use the Spectrum Analyzer (part of the Multimeter plug-in) to visually analyse the track's constituent components, as well as the band solo feature.

Now work your way through each band, adjusting the relative compression ratios, thresholds, and attack and release settings. On the whole, most compression ratios tend to fall below 3:1, with the most productive results achieved in the area between 1.5:1 and 2:1. Adjust your threshold to achieve the correct amount of gain reduction and dynamic control. Again, mastering is often distinguished by a lightness of touch, with typical amounts of gain reduction rarely exceeding 2–4 dB. Attack and Release times change the responsiveness of the compressor – avoid setting the release too fast (the compressor might start to pump) or squashing the transients too much with a fast attack. Finally, rebalance the bands using the Gain Make-up control. If you've compressed a band particularly hard, you might need to bring its level up to restore its position back into the mix.

9.7 Mastering in WaveBurner

For a fully professional result that meets the requirements of Red Book standards, you'll need to use WaveBurner. Preparing for this will simply involve rendering the files (ideally at 24-bit, without dither) ready to be imported into a WaveBurner session – after this, the rest of the mastering process (compression, editing, and so on) can be carried out in WaveBurner's domain. The advantages offered by WaveBurner stem from the fact that it is designed, from ground-up, as a tool for mastering – Logic, on the other hand, is a dedicated production tool. Track crossfades, for example, can be difficult to create in Logic – this is not the case in WaveBurner. Additionally, with support for CD-Text, UPC/EAN codes and index marks, WaveBurner is one of the most complete Red Book compliant applications available for the Mac.

After creating a New session (File > New), raw master files can be inserted in WaveBurner via the Region > Add Audio File menu, or by dragging the audio file from Finder. WaveBurner's screen is divided into four main areas, the most important being the Wave View and Overview areas (towards the top of the screen) that provide a graphic representation of the CD and the various regions that comprise it. The Region list, towards the bottom right-hand corner of the screen, lists the series of regions (or audio files) used in your session – a song, for example, could be constructed from several regions stuck together. The Track list, adjacent to the Region list, lists the Song marker used to break up the CD – more important, these can be completely arbitrarily placed, and aren't necessarily tied to the regions.

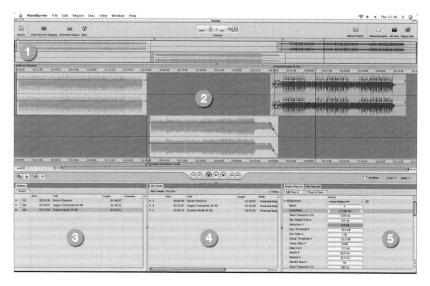

The WaveBurner interface:

1. Overview – presents a complete overview of all the regions currently used in your WaveBurner session.
2. Wave View area – gives a more detailed presentation of the regions, also where you can perform the various editing tasks (fades, trimming, and so on) in WaveBurner.
3. Region list – a list of audio files used in your session.
4. Track list – relates to the order and spacing of regions in the playlist and forms the structure of your CD.
5. Plug-in lists – displays the currently configured plug-ins used for the selected region, or the "global" plug-in inserted across the mix outputs.

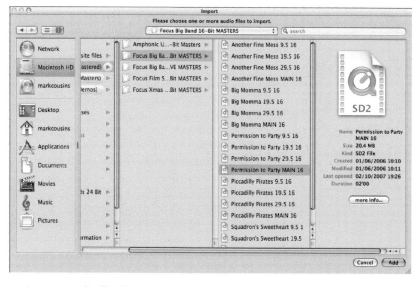

Importing new audio files that are ready for mastering.

9.8 Processing and editing

Audio processing is carried out in the Region and Mix plug-ins list. These work in much the same way as the channel and the master inserts did in the explanation of mastering in Logic. So, for example, the Region plug-ins can be used to process audio files imported into WaveBurner on a song-by-song basis, while the Mix plug-ins are applied to the entire programme output – in other words, all the tracks. Although the Mix plug-ins can be useful for processing en masse, they are best used for metering and audio analysis. WaveBurner includes the full set of Logic plug-ins, as well as support for third-party audio units.

With WaveBurner's specific focus on mastering, the editing tools work more quickly and more effectively than Logic's Sample Editor. Changing start and end points, for example, results in WaveBurner intelligently shuffling the other tracks in line with your edits. Fades can be quickly addressed by adjusting the nodes at either end of the region, with options to change the fade curve. More important, unlike trying to master in Logic Pro, these fades are carried out after the Region plug-ins, ensuring a consistent tone irrespective of the fades. Dragging regions over one another creates a crossfade – either to iron out edit points in the two track masters or to create smooth transitions between different tracks.

Plug-in focus 3

Limiting

Although Multipressor does a good job of bringing the overall dynamic range of your track more in-line with a commercial CD, it's still possible that your tracks will sound at least 4–6 dB quieter than a professional release. This is because fast transients (often from slightly louder drum hits) have slipped through the Multipressor's "net" and are still dictating the overall peak level of the track – even though the average level is somewhat below this. Professional mastering engineers will therefore make use of a fast-acting digital limiter (sometimes known as a Brickwall Limiter) to catch these peaks and reduce them. With the peaks reduced, the overall track level can be lifted to match the loud output currently favoured in mastering.

Logic and WaveBurner's Limiter both present a basic set of controls that should allow you to squash transients in no time. The main parameter, Gain, pushes the overall mix harder towards 0 dB – where loud signals would then conventionally distort logic outputs, the Limiter kicks-in and applies gain reduction (indicated in the top meter) to stop this from happening. In most cases, especially those that have already been through the Multipressor, you'll only need to apply about 3–4 dB of Gain, although you will need to listen carefully for any unwanted audio anomalies, especially distortion.

The Limiter plug-in provides that extra sense of loudness to your master, although you want to be careful how hard you push this addictive plug-in.

9.9 Dithering, bouncing and burning

By keeping your session masters in a high resolution, you will ensure the best possible signal integrity throughout the entire mastering process. Even so,

Plug-in focus 4

Other mastering tools

As well as the three big mastering tools (EQ, multiband compression and limiting), both WaveBurner and Logic also feature other plug-ins suitable for track sweetening. Stereo Spread, an effect often found on hardware mastering processors like TC Electronics' Finalizer, offers the ability to extend the width of a stereo signal – say, for example, the mix had been made slightly too mono. Unlike certain other Stereo Spreaders, Logic's mix widener avoids the use of phase in achieving extra width – instead it uses a frequency distribution system to pan alternating frequencies to the left and right-hand speaker, respectively. By using this frequency distribution method, Stereo Spread avoids the phase problems often associated when a width enhancer is put back into mono. In truth, its effectiveness is somewhat limited and it only works best with material possessing little or no stereo information.

Denoiser uses some clever trickery with FFT filtering to produce a cleaner audio signal in situations where the master possesses a large amount of unwanted noise. Like a conventional noise gate, Denoiser needs its Threshold

Although they have little everyday use, Logic and WaveBurner's other mastering plug-ins can be useful solutions to a range of mastering problems.

set careful to make best use of the effect – try locating a quiet section of the master, with the noise present, and set the Threshold just above this. Reduction defines the amount of Denoising taking place, although if pushed too hard the Denoising can sound almost as distracting as the noise itself! The Noise Type fader seems a little strange at first, but it does make sense – in the centre position, the fader is sensitive to noise across the entire audio spectrum, towards the top its bias is towards darker noise, and to the bottom it is more sensitive to high-frequency noise.

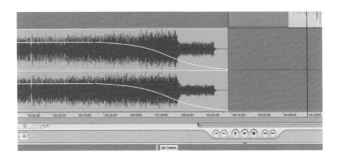

Creating fades in WaveBurner completely with a variety of fade curves. Drag the small nodes to define the length and shape of the fade.

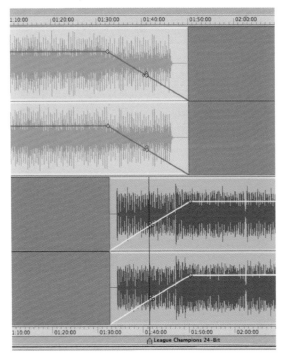

Crossfades are as dragging regions over one another

the reduction to 16-bit resolution and the application of dither are always inevitable. For such a vital process, however, it is easy to miss when and how WaveBurner applies dither. Technically, the controls for dither are actually part of the Bounce preferences (WaveBurner > Preferences > Bounce) – an option whereby WaveBurner creates finished files for the songs you've created (just like Logic's bounce). Bounces can be performed region by region (Region > Bounce Mix), or you can create a bounce of the entire CD as one contiguous "image" file (File > Bounce Project). Rendering the files isn't essential unless your computer's under too much strain with its DSP to perform a Burn, or because you want to move the files (with effects printed into the mix) into another application.

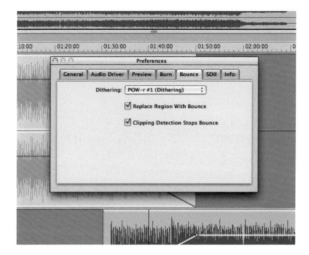

The bounce preferences contain WaveBurner's dithering options – applied either to a bounce or to the final burn.

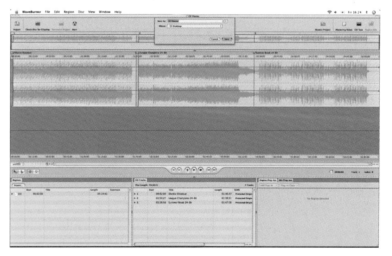

Use the bounce feature to render WaveBurner's output as an audio file, or to create a contiguous image file of the final CD.

What isn't immediately obvious is that the Bounce preferences also dictate how WaveBurner burns audio to a CD – particularly with respect to the application of Dither. Therefore, as long as the Bounce preference is set to dither, WaveBurner will automatically dither your files as part of the burn – without specifying the need to burn. In the rare case that you are importing files already processed and dithered, you'll need to remember to return to the preferences and turn the dithering off.

By pressing the "create disc icon" you can initiate the burn – although this is only the beginning of the process of learning mastering. For something so apparently simple (how difficult can two tracks and a multiband compressor be?), the real art of mastering professional CDs will take a lifetime to master. Ultimately, the art of mastering is a surprising blend of a gut instinct for what sounds right (in other words, the mastering engineers "ears"), alongside a complete understanding of the technical criteria of manufacturing an audio product. At least with Logic and WaveBurner you have some powerful tools in hand, and the opportunity to present your music in the best possible way.

Walkthrough 1

Editing and assembling a CD in WaveBurner

Step 1:

Start a new project in WaveBurner (File > New) and import your required 24-bit session masters (File > Import Audio File). Try resizing the three areas of WaveBurner's interface (Overview, Wave area, and Regions lists) to best suit your needs – here we've selected a small overview and Regions list, alongside a large, clear wave area for editing. With the songs imported, you'll probably need to perform some basic edits before going any further. To do this, click on the region you want to edit (from the overview) and click and drag either the start or the end point to resize the region. Leave some duration of silence at the start to allow for CD de-muting.

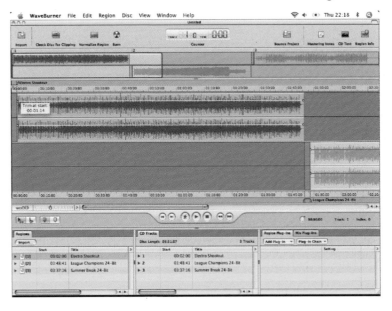

Step 2:

Add fades onto either the track's beginning or the end by dragging the small node at either end of the region. You can change the curve of the fade using the other two nodes that appear – sometimes the linear fade

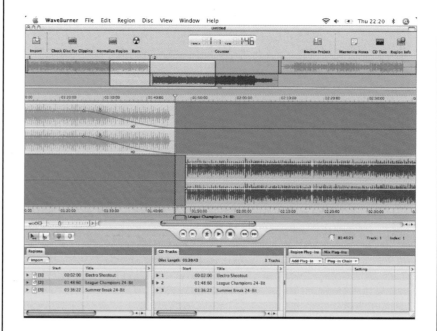

can be a little too obvious to the ears, whereas curves can be more gentle or extreme depending on the material you're playing with. Another important thing to look at are the gaps and pacing between the tracks, as illustrated by the purple-shaded "pause time" area. Try slipping the next region backwards or forwards to match the required pace of transition – this is an important way of setting the feel of the album.

Step 3:

Where a region is dragged back beyond the shaded pause area, a cross-fade can be created between the two tracks. As with the fade-in and fade-out, the curve of this can be adjusted to get the smoothest, most musical transition between the two tracks. WaveBurner will automatically place the Song marker halfway between this transition – if you want it placed elsewhere, drag the small purple marker on the top of the wave area back to the appropriate point. In addition to song markers, you can place index points – simply change the marker tool (bottom right) from purple to orange.

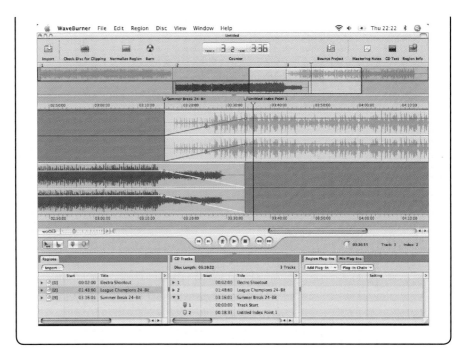

Walkthrough 2

Audio mastering and CD burning

Step 1:

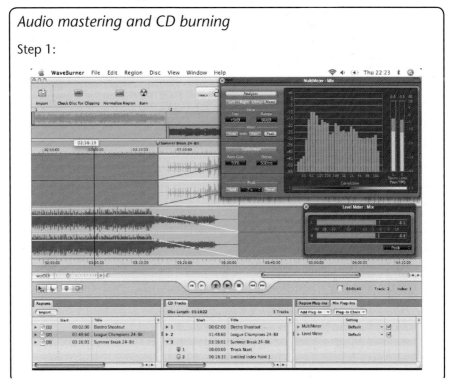

With the structure of the CD defined, you can turn your attention to the sound, in an attempt to get a "loud" consistent tone throughout. To perform this, some informative metering is essential so that you can hear and see detailed qualities of the track you're working with. In the Mix plug-ins list, insert an instance of WaveBurner's Multimeter and the Level Meter – this should give you a good overview of the timbre and the overall level of your CD. Try importing a commercially mastered track (similar to the CD you are working with) to get some reference to how things should look and sound.

Step 2:

Now turn your attention to the sound of the individual regions. Select a region from the Region list, click on the Region plug-ins tab, and insert the appropriate processing to achieve the desired sound. In most cases, a combination of the Linear-Phase EQ, Multipressor, and Limiter (in that order) should pull the mix up to the point where it sounds "loud and proud." Check the results against the other tracks so that you achieve a consistent tone. Most importantly, avoid excessive clipping or distortion, which you can identify through careful listening, or (from a technical perspective) using the Disc > Check Disc For Clipping menu option.

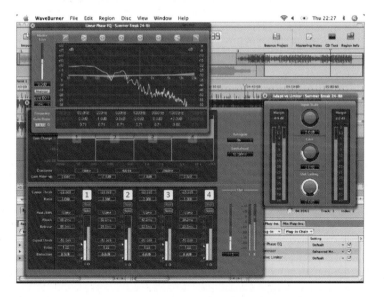

Step 3:

If everything's sounding sweet, you're ready to do the final preparation for the burn. If your source files are 24-bit you'll need to check the dithering status of WaveBurner. To check this, go to WaveBurner's preferences and select the Bounce tab. Now pick a dithering option from the pull-down menu – POW-r # 1, 2 or 3. Dithering is applied either when you Bounce files in WaveBurner, or as it creates the final burn. To set the burn in action, click on the Create Disc icon and insert a blank CD-R into your CD writer. To ensure a minimal error rate, use the lower burning speeds.

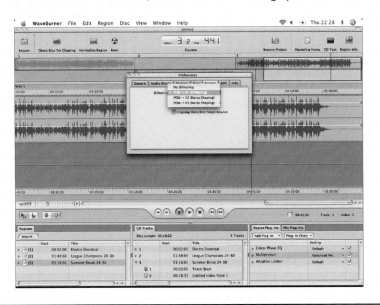

Knowledgebase 3

POW-r

POW-r comes with three options within Logic. One is labelled "Dithering" which employs a noise-shaping curve to reduce noise. The remaining settings offer two types of noise shaping. The Logic Manual describes the first noise-shaping algorithm which can "extend the dynamic range by 5 to 10 dB." The second noise-shaping algorithm is intended for work with speech as it can "extend the dynamic range by 20 dB within the 2–4 kHz range – the range the human ear is most sensitive to" (Apple Logic Manual). These noise-shaping curves have been developed to be sympathetic to the human ear and draw upon considerable research beginning with Fletcher and Munson's Equal Loudness Contours (More information on Equal Loudness Contours can be found in Rumsey, F. & McCormick, T. (2002) *Sound and Recording, An Introduction*. Focal Press).

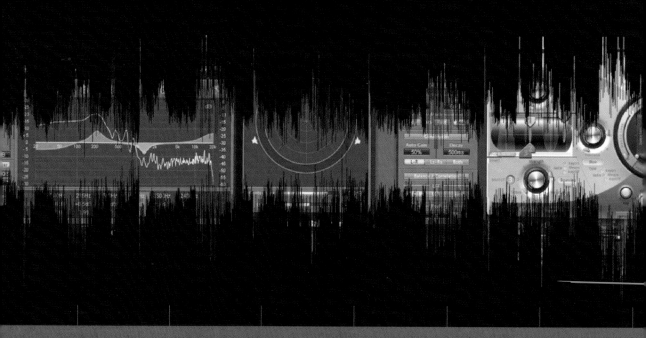

Logic and multimedia production

10.1 Introduction

Writing music to picture for film and television often required a whole host of equipment from some form of video player synchronized to your computer sequencer and Digital Audio Workstation. However, Logic takes this task in its stride, and today whole soundtracks can be written for picture from the one application. This has been made possible by the thorough integration of a visual player and high-level synchronization options from the ground up. This flexibility has meant that Logic has become one of the most popular products for film and television music production.

Additionally in the past few years, we have seen a sudden rise in other outlets for the music producer, such as gaming and internet multimedia. Logic with its integration of codecs to export audio as MP3, AAC and many other formats has placed it as a key production tool for many industries other than recording and production.

In this chapter, we take a movie file and begin integrating it into the Logic project and explore how to write effectively for film and television. We'll look at working with surround sound, markers, scoring, and the delivery formats expected from the industry.

10.2 Managing movies

Logic has been at the forefront of music and audio composition for picture for sometime and as such there are many ways you can work with visuals. You can, of course, synchronize Logic to an external Visual Editor such as Avid or Final Cut, if you so wish, but you're more likely to obtain a video file as an .avi or .mpg (Quicktime) ideally with embedded timecode so that you can work wholly within Logic itself.

Before starting work on the music and audio for your visuals, you will need to import the movie into Logic. There are a couple of methods to achieve this. The first is to go to Options > Movies. Here you will be greeted by another menu of options, which give the user great flexibility in integrating video into the Logic song. To begin to integrate movies within your song, go to Open Movie (Options > Movies > Open Movie). The second method involves selecting your movie using the Open Movie button in the Video lane within the Global Tracks, which is covered later.

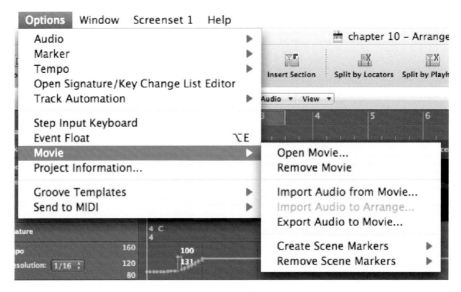

The movie menu is where main video files can be selected as well as Scene marker management.

Once you have located your movie in the browser, it will appear as a floating window which sits neatly on top of Logic. This can be moved around to fit the screen arrangement you have currently. This floating movie window can be resized to suit and this is achieved by Control+clicking the video window. This produces a new small selection box that reveals itself to offer you the choice of an alternative size for the window or the Video Project Settings.

Control+clicking the movie window opens up many options such as the size of the movie window itself.

The Video Project Settings allow you to control a number of things starting with where the movie presents itself. Within the Video Output menu it is possible to select the preferred output. This will naturally remain as "Window" which will open the movie in the standard floating window. For using a second monitor attached to the DVI output of your Mac, select Digital Cinema Desktop and then select the size of Video Format for that screen. This monitor cannot be used for any other function whilst Logic is in this mode. Other options include the professional DV output associated with the picture industry called DVCPRO HD and FireWire for sending to an external video interface.

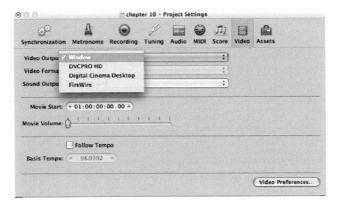

The Video Output options allow for most visual connections using a Mac.

The audio from the original movie can be piped through to the main output, although you may wish to control this by importing the audio into the Logic project, which we'll cover later. Other flexibilities are available here such as the movie start. For example, it is often unlikely that you'll want the video to start running from 00:00:00:00, but may indeed need some pre-roll for any introductory credits that might not have come to you yet.

If you're working on a small monitor or a laptop, Logic has a neat way of monitoring what is going on in your movie. At the top of the Inspector there will now be a movie header. Simply click on the triangle on the left-hand side to open up a small movie viewer.

If you're working on a laptop on the move, then Logic has a handy viewer installed in the inspector which will appear above the normal editable areas when expanded.

The Follow Tempo option fixes the speed of the video to the tempo of the Logic song. This link is set by something called the basis tempo. If the project tempo alters, then the speed of the video can relate to the difference between this and the basis tempo just like an Apple Loop would speed up to follow a new project tempo. To the bottom right of this Project Settings pane lies another button called Video Preferences.

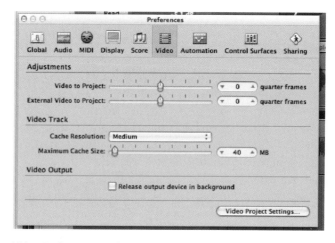

Video Preferences and project options offer precise adjustments for timing the Logic project to the movie amongst other things.

Launching the Video Preferences pane reveals some more settings that can be adjusted for our work with video. The uppermost sliders, Video to Project and External Video to Project, allow for the video and timeline to be adjusted against each other, thus altering the start point of the movie should perhaps some lip synchronization be slightly out from a provided dialogue track.

When working with video, Logic displays stills of your movie in Global Tracks (covered later). In order to allow Logic to do this, the computer needs to hold the movie in something called a Cache. A Cache is a form of buffer which holds some of the required data in memory for quick access. These settings allow you to specify both the Cache resolution of the video information and the maximum Cache size allowed. By altering these you can manage how much video detail you can store in the Cache and thus the response speed when scrubbing.

Extracting the movie audio content

Quite often you might be given a visual which has integrated audio, perhaps a dialogue track for the scene you are composing to. It will be prudent to extract this dialogue track for you to be able to manage its level and mute throughout the writing process. In some circumstances, for example, you may wish to mute the dialogue to ensure that you can hear your music properly before reintroducing it to see if it all fits together.

The quickest way to extract audio for this purpose is to select Extract Audio to Arrange (Options > Movies > Import Audio to Arrange). By selecting this, Logic goes to work immediately taking the audio from within the movie and moving the subsequent audio file to a track on the Arrange page. The other option is to import audio from movie which places the audio file in the audio bin. When the audio is extracted, its region will be timelocked to the position of the movie to ensure lip-synchronization.

10.3 Global Tracks

Global Tracks are an important feature of working to picture, not only for the viewing of video stills in the timeline, but also for the ability to change the tempo, time signature, and keep track of important markers in the project. Changing the pace of the music can be crucial when composing to picture and Global Tracks allow you to make these adjustments effectively as required.

Accessing Global Tracks is achieved by pressing "G" whilst on the Arrange page or clicking on the triangle to the left of the "Global Tracks" legend to the top left of the Arrange area. This should expand to show you a number of lanes that typically include markers, signature and tempo. To add to these, we need to choose to configure Global Tracks.

For example, a common way to view the visuals you're working with is to use Global Tracks. Video stills can be added to this timeline by going to View > Configure Global Tracks or pressing Alt+G. A list will present itself in which you can select the lanes you wish to see in Global Tracks. By selecting video, a new lane appears which can show a set of stills of your video against the timeline.

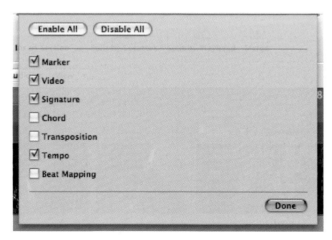

Whilst on the Arrange Page, simply press Alt+G to view the Global Tracks you'd like to view.

Within this lane there are some useful buttons. One of which is Open Movie, which offers you the opportunity to either select the traditional, larger floating video window if you already have a movie loaded. However, if you've not yet loaded a movie to the project, then a finder-like pane will appear to allow you to select a new movie.

Markers

As with the locators found on traditional multitrack tape machines, Logic can also offer the ability to return to any marked position on the timeline. This can be useful to return to any point of a project which is visited frequently when recording, such as the beginning of the Middle 8 or a Chorus. Additionally this can have some excellent uses when working to picture.

Logic terms these location points as markers and they can be seen in the time bar at the top of the Arrange area but are accessed in two main ways, either through the Global Tracks or via the Markers Tab in the List Pane. As you expand the Global Tracks, you will notice that the Markers Lane come into view. It is likely that this lane also needs to be expanded by pressing the triangle to the left of the track header. This lane shows any markers you may have set against the timeline.

To get instantly working with markers, simply place the playhead at the required position and then click the Create button. A new marker is created which appears to run straight to the end of the project, although there is no length specified. This will be interrupted by the next marker you create, or if you specify a marker length.

Markers can also be created on the fly whilst you are playing. This can be useful to note any aspects of a recording or take you wish to revisit when you listen back. By pressing Apple+Alt+Control and clicking on the Lane, a new marker is created whose name awaits you to edit. Simply rename the marker and press return. This can be really flexible as you label sections going through a song for example.

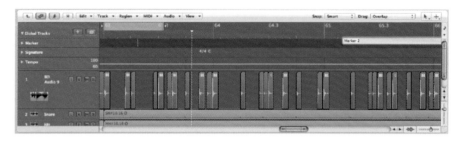

A new marker can be created and named by pressing Apple+Alt+Control and clicking in the marker lane.

Editing a marker can take place directly on the Global Track Lane by selecting the appropriate tool from the Click Tools Menu (Escape) then editing the marker. Using the Text Tool to name makers is very flexible and can assist your workflow whilst you play through the takes.

All markers are also reflected on the Marker Tab within the Lists Pane which can be found by pressing "K" whilst on the Arrange page. Alternatively click on the Lists button to the top right of the Arrange page. Next, select the

Marker Tab which shows all markers within your project in a long editable list. Here you'll see the markers arranged in a table with their position to the left-hand side and their Length to the right. Naming may appear the same as in the Global Tracks lane here but can be added to create a separate line within the Markers Lane which can reflect the part of the project. Simply double click the name of the marker within the tab to reveal the marker naming area.

Each marker is tagged as "Marker ##" or "Scene ##" where the "##" automatically numbers the markers in sequence. If you continue to write after this, then your text will be merged on the same line. But as in the example below, by pressing return, the Marker Lane can produce separate lane to show your description which can be very useful when working to picture.

Scene markers are different to that of traditional markers in that they are locked to the movie file using SMPTE timecode. They will not alter depending on tempo changes or edits in the project. It is possible to convert these to normal markers and vice versa by selecting the Marker Tab ("K") and selecting the appropriate Convert … command from the Options menu.

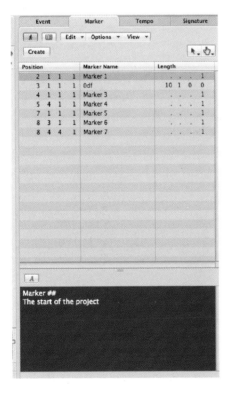

Markers can be renamed and fine tuned here in the Marker Tab. The renaming can also allow for more description within the Global Tracks.

Locating to the marker positions is simple and can be achieved firstly within the Marker Tab by using the Finger Tool from the Click Tools Menu (usually the Apple+Click Tool is set to this automatically) to click on any marker. The playhead will spring back to the marker point. To wrap the locators around the marker length press Option+Apple and Click. To access the start of a marker from the Global Tracks lane, simply press Option+Click a marker. Key commands can be programmed to respond to marker numbers if required, but are not part of the default key commands set.

Within the Global Tracks Marker Lane, an additional button called "From Region" can be seen. This feature allows you to select pertinent regions from which markers are to be made. Perhaps you have different folder tracks already cut to the length of your verse and chorus. In this instance, the markers would automatically follow suit and allow you quick access to these elements of the song.

The "Alternative" feature within the Global Tracks Marker Lane offers the opportunity to have up to nine different sets of markers. This might be very useful when working to picture where markers are required which represent the dialogue, others can be set in an alternative for working against the music and perhaps another for the visuals themselves.

Detect scene changes

It is important to get a feel for the structure of the visuals you are working with. A script or storyboard may not be provided and yet you will need to break the visuals down into useful, manageable chunks. The best way to get a handle on the visual material you are working with is to work scene by scene.

"Detect Cuts" is a button on the Video Global Track and stands for Detect Scene Cuts. Logic has a really useful feature in that it can detect scene changes in the video and produces markers for you to refer to or edit to suit. These take a little while to detect and will appear in the Marker Lane in Global Tracks as well as in the Markers List which can be found by clicking on the Lists icon on the top right-hand side of the Arrange page and then selecting the Marker Tab.

Scene markers are presented with a "film" icon within the list and are automatically labelled as "Scene 1, Scene 2, Scene 3, etc." Obviously, this process

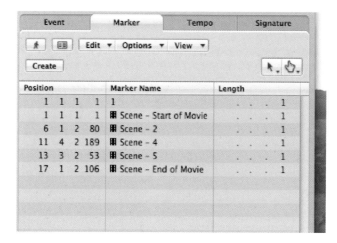

Scene markers are clearly seen within the Markers Tab and can be renamed and edited here.

may not be entirely 100% accurate, and it may be necessary to alter, add, or delete some of these scenes. To create extra markers to denote scenes that the process may have missed, simply create a new marker by pressing the Create button on the marker Global Track. This marker may need to be converted to a Scene marker as described above by opening the Markers Tab in the Lists area, then opening Options > Convert to Scene marker.

As you work through each of the Scene markers, you will need to move them to fit the movie more accurately. This can be easily achieved in the Global Track, where you are automatically given the resize tool. Alternatively visit the Marker Tab and edit using the SMPTE timecode.

Working with tempo and signature changes

Making changes to the pace of the music is often the key to writing effective music when writing to visuals. Whether it is to denote a new mood, or some dramatic news that has been announced within the dialogue, changing key and tempo, in conjunction with the accompanying visuals, could instantly move your audience to another emotional space.

Changing the tempo is very simple and can be achieved in the Global Tracks. Either select the pencil tool (Escape > 2) and click the tempo change at the desired place or choose the Tempo Tab from the Lists pane, which can be accessed by the icon on the right-hand side of the Arrange page.

Using the pencil tool can offer you a quick method to try ideas out in the project. Clicking the Global Track will reveal a new tempo which can be glided up or down to instigate a tempo change. You'll notice that the way in which these tempo changes are represented look very similar to that of the automation lanes and as such can be manipulated in similar ways.

For example, you might choose to introduce the piece at a slower tempo and increase it gently. To achieve this is easy, simply set the two different tempos at the ideal point. The jump will be instantaneous and will sound odd. There is a small point in blue at which the tempo changes from the original tempo to the new tempo. This first point can be moved backwards in time to introduce a more subtle curve to the tempo change, thus making it gradual.

The menus within the Tempo List Tab offer some more options to tempo management. For example, you can view the subtle steps that Logic has placed in the project to increase the tempo. To do this you need to click on the Display Details button to the left of the Edit menu in the Tempo Tab. This button contains a stream of 0's and 1's within it. Prior to selecting this feature, only the start and end point of the tempo change can be seen.

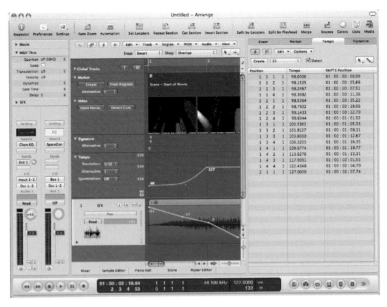

Tempo changes can be altered quickly and flexibly in the lists pane on the Arrange page or by editing the Tempo lane within Global Tracks.

With the steps expanded, it is possible to quickly edit any step to suit. As you edit the tempo of one of these discrete steps, the Global Track will reveal the steps graphically on screen. However, to create a more staggered, creative, or editable tempo change, simply choose Options > Tempo Operations, which can offer some more options to the on-screen Global Track.

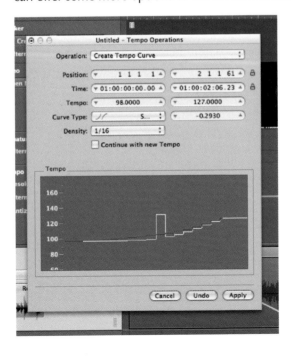

Tempo Operations dialogue gives some additional options when making tempo changes.

A new pane appears that allows you to edit more precise and in-depth tempo changes and alterations. There are many different editable features here such as the types of curves and their resolution. Within the Operation menu, there are a number of really useful features that could save time such as stretching the curve out, or simply scaling it a little.

The signature lane in the Global Tracks is split into two, the top is for the time signature and the other for the key. As with the Tempo lane, there are again two main methods of changing the signatures. The first is through the lists on the right-hand side Lists window, which contain a tab dedicated to signatures. Within this tab there are two buttons, one for time signatures (create signature) and the other to Create Key changes.

The other entry method is to click again on the lanes itself with the pencil tool (or simply double click with the pointer). Whether it is for a signature or a key change, a slightly different dialogue box greets you. The first of these is the time signature and this allows you to simply select the signature you want, but also gives you the opportunity to enable beat mapping, which we covered in Chapter 5.

Adding a key change in the Global Tracks is easy by double clicking in the lane where a dialogue box allows you to specify the change.

Adding a key change is really useful for quick ideas and changes and will work with Apple Loops and any audio you have recorded in this Logic song. The key signature here will enable the Loops to alter in real time to your changes. Simply select the point at which the key change is supposed to occur and double click (or click one using the pencil tool) and a dialogue box will emerge requesting the desired key and also whether you wish to disable double flats and sharps.

Alternatives

It is often difficult to be sure that the choices you make are definitely going to be right for the outcome of the visuals. Often it is nice to make changes and try out ideas without losing the original version. Logic has an interesting solution which allows different experiments.

Each Global Track has something called Alternatives, which offer different perspectives on the same track. Similar in nature to alternative playlists when quick swipe comping which we covered in Chapter 5, the Alternatives menu in the Global Tracks allow you to make some different decisions without losing your original root idea. Simply click on the Alternatives menu and choose an alternative number. It is useful to remember that by pressing Alt as you click on your new choice, the data are copied from one alternative to the other.

Alternatives are excellent to try out different permutations of ideas in Global Tracks.

Big displays

Working to picture is always dependent on timecode which governs the synchronization between different cameras and audio recorders for editing purposes. Timecode typically comes in the form of SMPTE (Society of Moving Picture and Television Engineers), which splits the time into divisions of Hours: Minutes:Seconds:Frames and in some cases sub-frames.

The transport bar always shows the SMPTE time of the playhead in the top left with the accompanying bars and beats below. When spotting sounds or effects to film, it will be probably necessary to see a larger timecode display. Control clicking the transport bar brings up a menu from which you can choose from the Big Bar Display or Big SMPTE Display. These are displays that remain within the darker section of the transport bar. There are also two options to show these as separate floating windows by selecting the Open Giant Bar Display and Open Giant SMPTE Display, respectively (shown below).

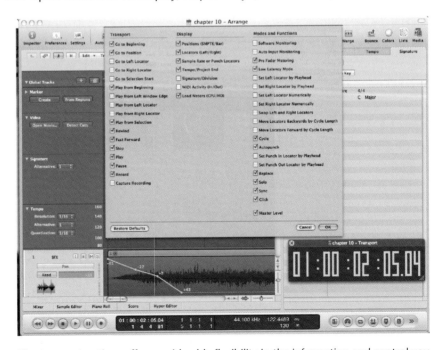

The transport options offer considerable flexibility in the information and control you have over your project.

Spotting audio

Working to picture will from time to time mean that you will be given a spotting sheet from your clients listing the sound effects. This will tell you where all the sounds need to occur to a particular timecode point and may also include an out time point. These points are given in SMPTE timecode and in Logic it is easy to edit to these points. It's worth sometimes working to a snap value of "Frames" when working to picture to ensure that each frame is synchronized with the audio.

The playhead's SMPTE position can be specified by double clicking the timecode readout in the transport bar.

Two main ways of spotting audio are on offer here. First is by using the Move to Playhead Position feature. Simply set the playhead to the point you wish for the audio to be placed. This can be best achieved by double clicking on the SMPTE display in the transport bar and typing is straight in. Pointing towards the audio region, click with control pressed down to reveal the contextual menu and within this will be the Move to Playhead Position command.

The more likely method is to edit using the Event Tab from the Lists window on the right-hand side of the Arrange area. Presuming you have called up all your audio regions onto the Arrange area, they should be listed here. To hone in on the parts you wish to spot, ensure that Link is on both the Events List and the Arrange page, and simply click on the region you wish to view. The Events List Tab is naturally set to show bars and beats and will need to be changed to view SMPTE through the View Menu (View > Event Position and Length in SMPTE Units).

To spot the audio, simply double click on the position column to type in the desired timecode. The audio should in theory be spotted, although it might be likely that you will need to delve in further to more accurately ensure synchronization with the picture. Nudging the regions by frames can be a really quick way of making things fit. As we mentioned before, it is often wise to work to frames for the time being, although there will be times where you wish to work in finer resolutions. To change the Nudge Value, right click on any region and select Set Nudge Value. Once set, simply select the region to be moved and either select the Nudge Left or Nudge Right from the region's contextual menu or press Alt+Cursor Left or Alt+Cursor Right.

A really fast way to spot audio is to position the playhead at the desired point and select the track you wish the audio to be placed. Next open the separate Audio Bin Window using Apple+9 or Windows > Audio Bin. Select the audio file in question pressing Apple as you click and the audio file will be placed at the appropriate place. If you then select another audio file in this way it will be butted up to the last audio file. There are a couple of useful key commands that can be brought in here called Pickup Clock (move Event to Playhead Position) and Pickup Clock and Select Next Event. These can be set in the key commands menu (Alt+K).

10.4 Synchronizing Logic

Working to picture or on larger projects might involve the need to link Logic to another player. For movie work in the past, it had been often the case that the video would be on another machine, perhaps tape-based and Logic would need to synchronize to it. Similarly there may be times when Logic needs to synchronize to another multitrack system, whether that be a legacy open reel 2-inch 24-track tape machine such as a Studer A800, Otari MTR90, or simply another ProTools rig on another computer. Either way Logic has a suite of features to cope with your synchronization needs.

Logic syncs in a number of different ways that are industry standard. The main system used for video is something called SMPTE (Society of Motion Picture and Television Engineers) and is an audible code that can also be translated into a digital equivalent used in MIDI called MIDI Time Code (MTC). MTC is a popular system used between MIDI devices and in the latter example above, it is likely that this would be the protocol used to connect a ProTools rig to your Mac with Logic on.

In the either case, you will require a device that understands the time code protocol. All MIDI interfaces should understand MTC, but only some devices are able to translate SMPTE Time Code to MTC such as MOTU's MIDI Time Piece.

Getting locked

To tell Logic that it will need to slave to another time code signal requires you to select the customize transport bar preferences by Control+clicking the transport bar. Select the Sync check box which will generate a button on the transport bar which looks like a clock with a large arrow facing into it.

The sync button seen here in blue can be selected from the customize transport bar options. With this selected, Logic works in slave mode to an incoming synchronization signal.

In an ideal world, the synchronization should be as simple as selecting this button, and wait for an incoming sync signal to get the project moving. However, invariably you will need to edit the synchronization settings to suit the project you're working on. To do this either Control + click the sync button on the transport bar and select synchronization settings or vising File > Project Settings > Synchronization.

Right clicking the sync button on the transport bar gives you very quick access to the pertinent settings such as synchronization source and settings.

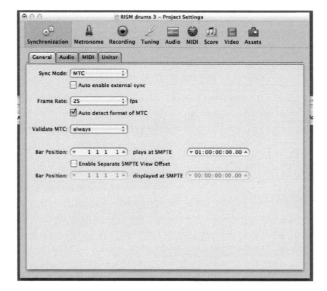

Use the synchronization settings to select frame rates and useful offsets.

Within this pane lie a number of key features for working with external synchronization sources. The frame rate of the synchronization is very important. Historically frame rates differed depending on where in the world you worked, so for example 30 frames per second for television in the United States of America, whereas typically in the United Kingdom, we worked with 25 frames per second. This was a simple division on the frequency of the electricity supplies in these regions at 60 and 50 Hz, respectively.

Strictly speaking we're no longer governed in quite the same way and Logic can work at whatever frame rate is required. Within this menu are considerably

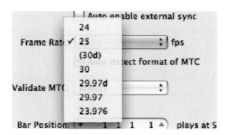

The frame rate will need to be matched to your synchronization source to keep up!

more frame rates than we've hinted to above and it will all depend on the project you're working on as to what frame rate you'll need. Safe to say Logic can handle them all!

Within the synchronization settings pane are some opportunities to offset the time code to the arrangement. These offsets are very important when working with film. For example, you may wish to use a fresh Logic project for a Scene 1 hour into a film. It would be rather silly to start the Logic project at 1 hour also because things like Freeze tracks would take a considerable amount of time to process!

Setting the offset is easy. Simply select the SMPTE time at which your song should play. For example, Logic will open with Bar 1 playing at 01:00:00:00:00 which means 1 hour. SMPTE reads as Hours:Minutes:Seconds:Frames:Sub-Frames. However, there will come a time when you will need to alter this to suit. In the example above where we'd like to start the new Logic project at 1 hour into the movie, we'd need to set an offset of an extra hour. As such we'd want the SMPTE reading here to read 02:00:00:00:00. Logic will then offset the incoming time code to allow Logic to start playing at 2 hours.

Below the main offset selection is an option to allow the bar positions to show the absolute time code reference, which can be useful when working to external sources. It is important for example that the time line you work to reads the same as the incoming timecode. Due to the offset in the example above, the Logic project will show a time of 00:00:00:00:00 despite being an hour into the film. This can be changed here so the project starts Bar 1 at 1 hour, but also shows an accurate portrayal of the time code it is receiving.

Being the master

Logic also has the ability to be the master and to allow other devices to slave to its time code. In this instance, it is necessary to return to the synchronization settings as outlined above and click on the MIDI tab. Here we can see a checkbox titled Transmit MIDI Clock, which refers to the tempo settings of the project. This clock pulse can be outputted using the MIDI protocol to connect other devices such as effects units whose delay settings can respond in time with the project. Alternatively in rare occasions this can be used in part for synchronization.

Below this is the Transmit MTC checkbox and output menu. Here it is possible to select a device (or "All") to send the MTC to. This will enable any device connected using MTC will follow Logic. For example, if the ProTools rig was to sync as a slave to Logic, then simply connect a MIDI cable between the output of Logic's interface to the other computer with ProTools' MIDI In.

MIDI Machine Control is a remote control protocol that enables connected equipment to be controlled by the another device. For example, Play on the master could be engaged by controls on a remote slaved device. This can still be a hugely beneficial option to employ across large studios.

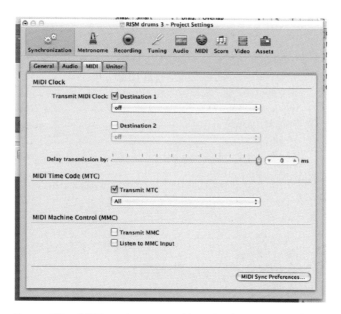

Transmitting MTC can be engaged by selecting an output from the Transmit MTC menu.

10.5 Score editing and music preparation

Although the Score editor can be used for MIDI editing activities in Logic, its most useful features lie in the "preparation" of music – in other words, taking the raw MIDI information that you've performed in the project and transcribing this as finished, musical notation. To be fair, though, Logic doesn't compete with the publishing-standard output of dedicated scoring programs like Finale or Sibelius, but its features are more than adequate either to produce parts for a small-scale overdub session, for example, or in the case of a full orchestral session, a sensible "intermediate" format to present to a proper orchestrator.

The process that we're going to explore here, therefore, is the rudiments of taking an existing MIDI composition and turning that both into a full score for the conductor, and parts for the musicians. Rather than being a complete exploration of the scoring features and the art of orchestration (which is a book in its own right!), we're going to take a look at the essential processes and steps that guarantee readable, usable results in the shortest amount of time. If you want to produce even more effective scores, though, it's well worth exploring further to see just how effective the Score editor can be in this task.

Preparing your MIDI files

As the old adage goes – garbage in, garbage out – and this is never more true than in the process of creating a score. Although it's easy enough to open any region in the Score editor so as to see a notation-based view of the music, it doesn't guarantee that the score is legible or playable. Arguably the clearest example of this is strings. In your MIDI arrangement, it's highly likely that the entirety of the violins, violas and cellos, for example, will be amassed to one generic string patch. Although this polyphonic approach makes sense for MIDI production, it isn't how players expect to read a score, as in reality, each instrumentalist will be expect to be presented with a single line, with the full score combining all the single lines onto a page.

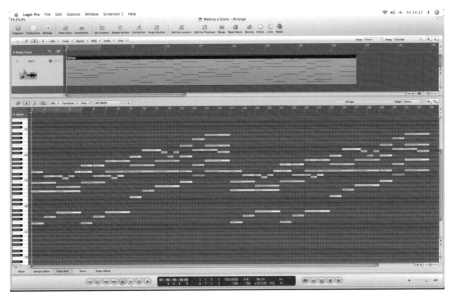

Here's a typical "massed" string part, with all the violins, violas and cellos being triggered from one generic sound.

The first step to prepare an effective score is to divide the music into a number of distinct lines (one for each player, or part) within your Logic arrangement. As we saw in the MIDI sequencing and instrument plug-ins chapter, there are a number of ways of doing this – from simply copying the part over and deleting notes in turn to features like Functions > Note Events > Voices to channels, followed by Split/Demix > Demix by Event Channel. Either way, you should end up with a series of regions for each part, each of which should be assigned to a unique named track for each instrument you want to appear in your score (like violin I, violin II, viola and cello, for example).

With the parts split, you might also want to glance through and double check any timing issues, as well as the precise duration of notes. For example, in a

MIDI arrangement, it's easy to have notes finishing 1/16th before the end of a bar, but this can lead to some strange looking note durations in the score, alongside unwanted rests. For simplicity, it's also worth merging each part into a single region lasting the full length of the score, unless you want parts dropping in and out of the finished notation.

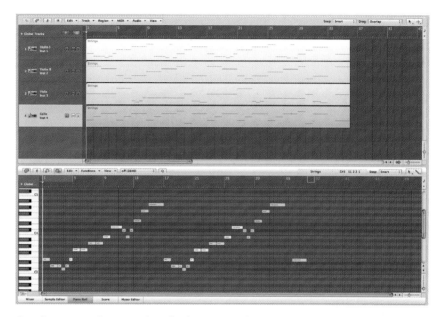

Creating a score from a series of split parts, each assigned to a different named track, will produce a more legible overall score, as well as facilitating the printing of individual parts.

Opening the Score editor

With the MIDI data prepared, you can now go and open the required parts in the Score editor, by drag-enclosing the required instruments and opening the Score editor via the Window menu. Although, of course, you can also open the Score editor directly from the arrange window, it probably makes most sense to have the complete screen dominated by the scoring features.

On opening the Score editor, you should have something on screen that approximates music, although as with the raw MIDI data, this will need to be tweaked to make the notation as legible as possible. For example, most of the instruments will probably default to a piano stave (with both a treble clef and bass clef on the same line!), and it's probably the case that the music simply runs across the full width of the screen, rather than being displayed on a page-by-page basis.

The first change, therefore, should be to turn the Score editor from its standard view into a page view, via View > Page View. Next comes the task of setting

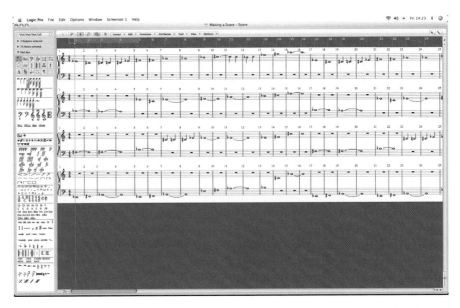

On opening the Score editor you'll be presented with basic score, although this will need to be modified to make readable.

the staves. In keeping with the workflow on the arrange window, you'll find that each line of score has its own set of score parameters, available via the inspector to the left-hand side of the Score editor window (note that this might be minimized at first, or even with the inspector itself hidden). By clicking on each line of the score, therefore, you should be able to select an appropriate style for each part – maybe selecting treble for the violins or the bass clef for a cello part. Note that for any transposing instruments (like Horns in F), Logic will automatically transpose the "scored" version of the instrument, although the "played" version remains untouched.

Adding expression and score markings

To add interest to the score, it is vitally important to add expression markings, otherwise you end up with a sterile and lifeless recording. To help with this, the Score editor provides a part box, which allows you to place a series of expression markings onto your score. These could be as simple as pp, mf, or ff dynamic markings, right up to trills, or crescendos and diminuendos. What you should notice is that the score stylings are attached to a particular part or line in question (the corresponding part will be highlighted blue when you drop the object), so that even when we move back down to part level (rather than the full score) these markings will be carried with the part.

Alongside expression marks, the part box also allows you to set the key of the piece (drag this to the first bar of the score, and then any subsequent bar should a key change occur), as well as text-based objects like the song name.

Adding expression markings, and so on, will make your
score both clearer and more expressive.

Shrinking to fit: score settings and score sets

Changing the broader qualities and appearance of the score can be done via
the score tab of the project's settings, also assessable via the Layout menu.
Although most of these settings (margin spacing, and so on) will be best left
in their default state, there are a number of useful functions. For example, the
Number and Names tab allows you to change how instrument names are displayed, either using their Full Names throughout the entire score, for example,
on an abbreviated Short Name for pages after the first page.

Another part of the full score's appearance is the percentage scaling of the
staves on the page. For example, on a full score it's often important to see as
much of the music as possible, so as to avoid excessive page turning, as well
visualizing the entirety of the arrangement at any point in time. This is of particular relevance to a full orchestral score, where you might need 30–40 different staves on the page at any one time.

Technically, when we selected our particular group of regions we created what
is known as a score set. As the name suggests, a score set assembles a group
of instruments that will appear together on the score at the same time, theoretically letting you omit certain tracks on the arrange from being displayed in
the finished score. In addition to this, you can also define a percentage scale

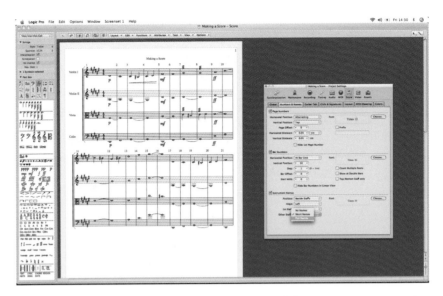

The score tab under the project settings allows you to define some macro properties about how the score is laid out and printed.

of a score set, allowing you to shrink or expand the score's size accordingly. Go to Layout > Score Sets... to access the score sets window, which you'll need to ensure has its "local" inspector open in the window so as to access the Scale [%] parameter.

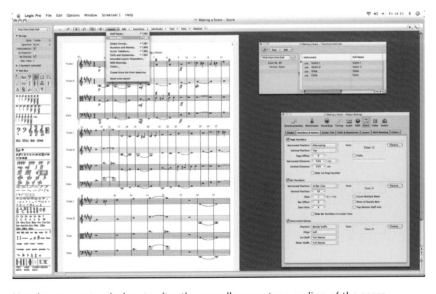

Use the score sets window to alter the overall percentage scaling of the score.

Printing the parts

To print off individual parts from the score, try double clicking on the stave in question. This should drop you down one level (the equivalent of going into a folder in the arrange), allowing you to see a single part with its associated score markings. So as to make the part clear and legible, you might want to consider raising the Scale [%] parameter as part of the score sets controls. Remember, though, that with both parts and the full score, you have the option to print to a PDF (via the OS) so as to create a saveable "hard" copy of the score for future reference, printing, or indeed, as an email attachment for your music copyist.

10.6 Surround sound in logic

Surround sound, whilst synonymous with the film industry offering that extra dimension of realism, is now becoming popular in some circles for music production with many artists' back catalogues being mixed for the medium. Whether working to picture or in music production, you'll inevitably find yourself being expected to create a surround sound mix at some point. Fortunately, Logic has a great deal of expertise in offering surround sound with the inclusion of surround sound panners on each channel and a 5.1 Reverb PlugIn amongst others.

To work with surround sound, in this example 5.1, it is important to ensure that you have the speakers you require to monitor each output. You'll need your main stereo pair of monitors, plus a centre speaker, two surrounds (left and right) plus a subwoofer ideally. Connecting these up will usually mean a direct connection from your audio interface or desk to the amplifiers or powered monitors. The connections you make are governed by the Surround Preferences (Logic Pro > Preferences > Audio > Surround).

Within this pane are three tabs relating to the output, the bounce extensions, and the input. The input arrangement should follow logically the output assignments and is used for making 5.1 recordings. The bounce extensions tab simply refers to the filename extension which is important when preparing for mastering. Concentrating on the output tab, you can select how Logic shows its outputs. Typically this will default to Logic's own interpretation of 5.1. There are options for ITU's and WG-4's output arrangements. The International Telecommunications Union's is standard for 5.1 surround sound for most professionals. However, the WG-4 standard, which can be selected by the button below, is the choice of the DVD forum. At this stage, it is just important that the right outputs are connected to the correct speakers for accurate monitoring.

Getting started

To get started with surround sound in Logic you could either select a surround sound template from the chooser, or you could alter your current stereo

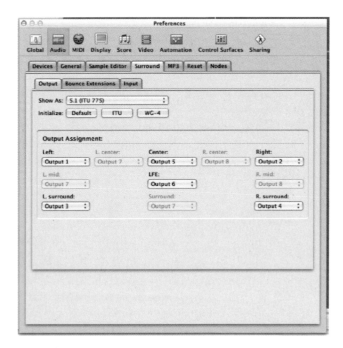

Selecting which audio outputs from your interface are sent to which speakers can be an important part of setting up your surround sound studio.

production to work in surround. Choosing Logic's own template offers 24 surround ready channel strips all bussed to groups including a 5.1 incarnation of Space Designer. This is excellent for that 5.1 project you're about to start.

However, there will be times when you'll need to alter your session to output in surround. To do this, simply choose the channel strips you wish to pan in surround, then visit the output assignment button. The drop down menu should now show the possible output assignments, the last of which is Surround. Click this and notice the master channel strip alter from stereo containing two peak meters, to a surround fader with five meters.

Working with Surround is very similar operationally to that of stereo except panning is different. You will have noticed a circular panner on each channel. When double clicked this opens up the Surround Balancer pane which shows a good sized panner showing a visual representation of the space with the five speakers facing inwards. To pan a mono signal, simply drag around the ball within the space to inform Logic where you'd like it to come from. Logic will calculate how much signal should come from each speaker.

For stereo signals, Logic's Surround Balancer shows three balls. The "L" and "R" balls represent the two discrete mono signals that go to make up the stereo signal. These balls are linked and allow for you to consider how wide you want the stereo signal to go within the surround space. The third ball is the overall pan that will alter the direction the stereo image comes from. In the larger surround panner on p. 326 the stereo signal is shown biased to the right.

Other controls on the Surround Balancer include the centre level and the LFE level. The centre level refers to the image created by the centre speaker. As surround sound is derived from the film industry, the centre speaker has traditionally been placed behind or around the projection screen and due to its focal point it contained all the dialogue. As such the music was usually omitted out of this speaker, leaving it to create the phantom centre image it currently creates with stereo reproduction. With a centre level control in Logic, it is possible to decide whether to allow any signal to come through the centre speaker at all.

Panning stereo signals in surround can make the process of mixing somewhat complicated. With two speakers, it is clear where the stereo signals go and how they might be managed as they can only come from within the space provided between the two speakers.

Surround sound in its popular incarnation means that there are five speakers through which the stereo signal can be routed. Does this stereo signal simply go to the left- and right-hand speakers and faithfully represent itself, or is it reinterpreted for surround and manoeuvered around the sound space? In the past, a stereo pair on a guitar might have stayed centralized to the centre, but with a multichannel set up it is possible to say rotate this 90° so that the left-hand channel could be reproduced from the right monitor and the original right channel from the rear right monitor, hence keeping the stereo image but shifting it in the sound space as shown in the example below.

In this view, the panner can be seen managing a stereo signal. When panning stereo source, there are extra controls to change the separation of the signal.

Three further controls are available from the bottom of the Surround Balancer. To obtain access to these, simply click on the small triangle to the bottom left of the pane. Three sliders appear that offer you the ability to alter the separation of the panner. In the diagram above, the orange square represents the bounds of the possible separation of the panners. However, in some examples it might be necessary to reduce this to reproduce a less wide image of the original. The three sliders correspond to the front left and right image (Separation XF), rear left and right image (Separation XR), and separation between front and rear (Separation Y).

In this example, the separation has been reduced to limit the width of the image. Also it is possible to see circular lines spread around the circumference which indicate which speakers are receiving a signal from which channel.

The LFE level refers to the Low Frequency Effect, sometimes also called the Low Frequency Extension, which is the "0.1" of 5.1. This is what is commonly known in the industry as a subwoofer. The LFE level control is placed within each panner as it is often undesirable to send everything to the subwoofer, when the main monitors should handle most bass information perfectly well. This again has come from the film industry where the LFE is used to reproduce rumbles, explosions and impacts. The 5.1 standard means that all the main five monitors should be full-frequency and should reproduce bass well for most applications. Hence, the LFE is included to add weight to the bass end. However it is worth remembering when mixing that for many audio applications, including home cinema systems, the subwoofer is not included as an "effect" but as the only bass generating driver, leaving the satellites to consider only mid- to high-frequency ranges.

Knowledgebase

Surround Sound Plug-Ins

Working in surround sound requires that you think slightly differently when working with effects such as reverberation. You could employ a stereo plug-in or you could choose to use two in tandem for front and rear if need be. Logic has fortunately thought of this and produced an intuitive set of surround ready plug-ins to get you instantly working.

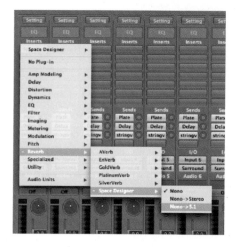

Choosing a surround plug-in can improve your 5.1 mix. Note how a mono track will be processed with a surround sound plug-in thus changing its output to 5.1.

These dedicated plug-ins include the renowned Space Designer whose 5.1 algorithms are superb. Other 5.1 dedicated plug-ins include: Delay Designer, Chorus, Tremelo, Flanger, Microphaser and Modulation Delay.

The plug-in behaves in exactly the same way as the stereo version, but the response is returned in 5.1 as per the original impulse. To place a surround plug-in in your mix can be simply achieved by selecting it from the menu. Most 5.1 plug-ins can be applied to a mono track whose output will be

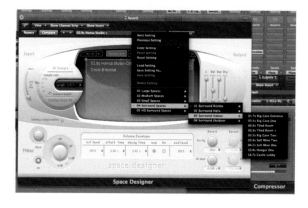

Space Designer comes with some impressive, dedicated surround algorithms.

converted to surround. However these processes are best added to a 5.1 bus or output.

Other plug-ins do not require specific surround algorithms to operate with multi channel information. These will therefore work in "Multi Mono" mode for processing multi channel information. Plug-ins such as the compressor, which utilizes this mode, will show some banks using additional switches located above the main controls that are not found on its stereo or mono counterpart. The Configure switch allows you to configure which compressor bank is attached to which outputs. Therefore it is possible to strap a compressor across the front speakers, another to the rears and one dedicated to the LFE channel. Alternatively you might wish to alter this so that the Centre channel has a dedicated bank for use in film work where the dialogue is of paramount importance.

Plug-ins which do not require specific surround algorithms work in Multi Mode such as the compressor. Note the selector switches above the main plug in to select which compressor bank operates which outputs.

10.7 Delivery formats

Bouncing to Stereo and Surround

At some point it will be necessary to present a mixed output of your production. If you are working within Logic, then it will be necessary to do a bounce and depending on the application, different choices will need to be made. One of these choices will be of course whether to produce a stereo or a surround bounce. Bouncing surround mixes is achieved in the same way as traditional stereo mixes.

Once your surround sound mix is balanced and ready, simply select the "Bnce" button from the master channel strip in Logic's Mixer. A new dialogue button will emerge showing a number of options for editing. The first to note is the destination for the files once you have bounced them. Choosing the arrow to the top right will open up this dialogue to show more of the file structure of the computer.

Bouncing to surround is as simple as clicking the "Bnce" button on the Master Fader Strip. It is important to select the surround bounce checkbox and choose whether you wish the file to be interleaved or separated out.

Moving on to the left-hand side the Destination table can be seen which allows you to configure the file type you wish to export to. For most applications this is likely to be PCM, which means unaltered or data compressed information, unlike the MP3, M4A options below it. The DVD-A option is an excellent option for surround sound as it will create a disc which can be played in most DVD players connected to a surround sound system.

Below this are the range controls allowing you to specify the start and end points of your bounce. This might be useful in the instance that you might need to export each scene separately for a dubbing mixer. Additional features include whether you wish to monitor the bounce in real time, or allow the computer to manage the event offline. You are also given the opportunity to automatically normalize your audio as it is bounced.

The right-hand side of the dialogue box includes a number of drop down menus that relate to the file destination type you wish to export. First up is the option to choose the file type. There are four choices here starting with Sound Designer II. The SDII format was created by Digidesign and has remained a standard for many years. Its benefit is that it is generally timestamped meaning that it can be moved and can be recalled to its original position at any time. This can be extremely useful when working to film for example or when transferring session between DAWs which accept the SDII format. The second incarnation is the Audio Interchange File Format (AIFF) which is generally the common format used on the Apple Mac. This file format is not subject to any data compression and is generally an accepted file format for use between software and the industry. AIFF does not contain timestamp information. Next is the Wave format, or .wav and is perhaps the most common of audio file formats for exporting to another computer for further mixing or mastering. Finally in the menu is the Core Audio File (CAF) which is an all encompassing format which contains a wide variety of different formats in one and is common for Apple Loops.

The next drop down menu is labelled Resolution and relates to the bit depth you wish for your bounce. This can be set typically at 8, 16 or 24 bits. Following this is the sample rate selector where a wide variety can be chosen from for a wide variety of applications which will be determined by your destination. The File Type box allows you to set whether you wish for the files to be interleaved together or split files for each channel.

Dithering is required in situations where a lower resolution bounce is required than the project's set resolution, for example when a 24-bit recording needs to be dithered down to produce a 16-bit CD ready file. Dither is a method by which the waveform is more accurately encoded for 16-bit reproduction and has been discussed previously.

The next, and perhaps most important feature for delivery of a multichannel export, is the Surround Bounce checkbox. This needs to be checked to enable that each channel of the surround field is exported as a separate file. The extensions for which are changed in the Audio Preferences as described before.

Exporting to DVD

Logic now allows you to export a DVD-Audio ready disc directly from the bounce dialogue. Within this window, select "Burn: DVD-A" and the right-hand

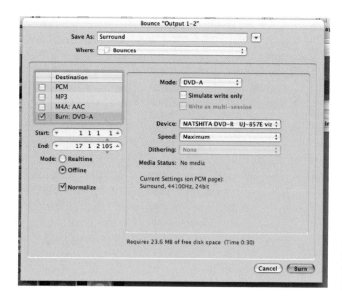

Burning a DVD-A can be achieved straight from Logic's own bounce window.

pane should alter to show you the Mode. There are two choices here. Either DVD-A which will produce a 5.1 mix for use in DVD-A players, whilst the other "CDDA" allows for a high definition (HD) stereo bounce of 24 bits and a 192 kHz sample rate.

For working to DVD Video, surround sound files need to be compressed using Dolby Digital Professional (AC-3) format. To achieve this, we need to use Apple's Compressor software which is part of Logic Studio or similar encoder software.

First we need to bounce the Surround Sound mix as split files, meaning that each stream (Left, Centre, Right, Left Surround, Right Surround and LFE) is an audio file in its own right. This will then allow us to indicate in Compressor what files are for which output. Before bouncing, it might be worth checking the bounce extensions found in the audio preferences pane as discussed earlier.

Exporting video with audio (dubbing)

Either during or at the end of the writing process for the film and television visual, you will be asked to send a copy with both the video and audio together. This process is known in the industry as dubbing and is something that is normally handled by large post-production houses, but for demo purposes, you can merge the two files together by selecting export audio to movie from the movies menu (Options > Movies > Export Audio to Movie).

A new dialogue box entitled Sound Setting will appear that allows you to choose the nature of the audio file. It is here that you decide upon the data compression format, sample rate, bit rate, and whether it should be stereo or mono. These settings are useful as a director or production company may wish to see your work in progress and the best vehicle to deliver this to them is over the internet via email (if small) or File Transfer Protocol (FTP). It is, therefore, handy that the compression formats allow for a wide variety of standards to suit each and every client.

Once these settings are decided upon, click OK and choose the location for your final files. It is worth creating a folder at this point to ensure quick and tidy file management as there may be more than one file given the compression format you have chosen. Next click Save and wait whilst Logic compiles your video for you. Once completed, this file should be readable by Quicktime.

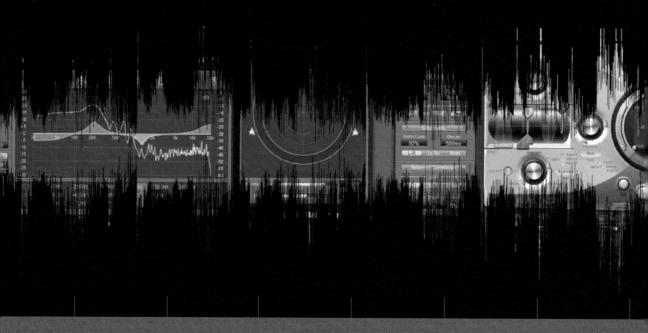

Optimizing Logic

11.1 Introduction

Working with any music software will expect you to work within its structure and design. Logic is part of that pack, but offers the user a number of options to personalize the way in which you work with the application. One of the most exciting things about Logic is its flexibility and configurability to the way in which you need and like to work. Music production requires a fluid work-flow that moulds to your way of making music. Doing this requires some initial thought and preparation, but it will improve not only the results but also the speed at which you work.

Logic has developed over the years an arsenal of features, tools and techniques to help the workflow of productions make the best use of the computer's interface. In this chapter we'll explore some of those features and how to tweak Logic to fit your workflow.

11.2 Templates

As you launch Logic, you cannot fail to notice the chooser which allows you to get started with one of the program's in-built templates. Many of these are very workable and are an excellent starting point. However as you begin to work in a particular way, you'll want your settings just so, thus saving your time and allowing you to be more creative. It is therefore worth preparing a template of your own with all the connections, plug-ins, windows and other features ready to go to work.

It is best to spend some time thinking about the way this template will look and function. Consider the way in which you use Logic before setting up the ultimate template. You may need to set up different templates for doing different projects in a wide variety of genres. To do this load up the Empty Project template from the "Explore" collection in the Template Chooser. Alternatively you may just wish to alter a template from the chooser which suits the way you work with your personal samples already preloaded into EXS24. Add to this project, your external MIDI devices you use frequently and give them tracks, your preferred organization of Screensets, and then choose to save as a template (File > Save as Template...).

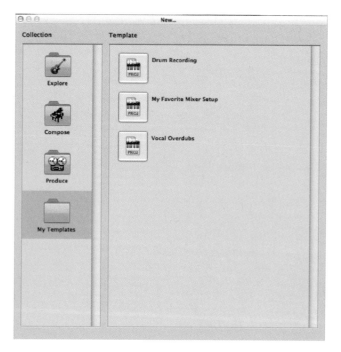

Logic's template chooser can save time when starting a project using a template of your own.

For example, if you do a lot of drum recording with a guide bass, guitar, and vocals, it might be prudent to set up a template for this kind of recording, and perhaps another for working to picture. The benefits here are that all the tracks will be pre-labelled, and therefore your audio files too when you hit record. This can be extended further to consider the mix, and if you find yourself using the same plug-ins and settings for your components of the drums, the template can have these pre-loaded as in the diagram opposite. It is perhaps sensible to put these into bypass mode (Alt + Click) as this can save on DSP power. In this example, the drum outputs are all set to go to Bus 3, which in this case has been set up as a Drum Group or Stem.

Once you have created and saved your templates, they will appear in a new collection Folder within the Template Chooser called "My Templates." These should provide you with personalized and honed platforms upon which move forward fast with your projects.

11.3 Screensets and windows

Until this latest version of Logic, there were many windows to contend with. Each editor, mixer, list and the Arrange Window were all separate and individually

Taking a look at the screenshot above, it is clear that a simple template such as this can really assist you in your preparation, thus saving lots of your time when it comes to record. Just set the levels and press "R." Having your commonly used plug-ins ready to rock really saves all that time flowing through different menus on each strip.

controllable windows. Hard to believe now with the redesigned Arrange window with its 'access all areas' philosophy. Logic's solution to managing all those errant windows remains and is still very relevant in certain circumstances.

Working in Logic may require that a significant number of windows be opened on top of each other at a time. The most obvious examples would be an Arrange window and an editor open at the same time, or the Arrange window and mixer together. Clearly the new Arrange window can deal with most of these adequately. However, if you wish to open other windows, most applications would use the Apple + "`" shortcut to allow you to toggle through the open windows in the program one by one. However, given the need to access so many different combinations of the Arrange window, editors, mixers, etc., Logic adopts what it called Screensets to allow for quick and easy navigation.

Screensets are Logic's ability to change what windows are viewed on the screen, or screens at any time. On the title bar of Logic to the right-hand side of the Screenset menu you will notice a number. This indicates the Screenset you have chosen, and pressing one of the numbers on the numeric keypad can change this, or you can click on the Screenset menu for more options. Within this menu you have the opportunity to rename your screensets, dupli-cate them, and delete them. The key additional features available within this menu are "Lock" and "Revert To Saved."

Lock allows you set the way in which the screen looks so that it cannot be altered. A small bullet dot will appear to the left of the screenset number to indicate this. This can be really useful when you need to rely on your screensets being just so! There will be times when the screenset changes you have made don't work for you and you wish to revert to your original decisions without losing the audio recording and editing you have done. To do this choose the Revert to Saved option from the Screensets menu.

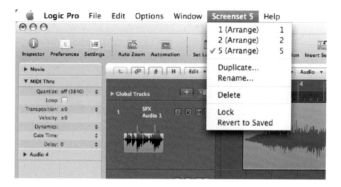

The screenset menu not only gives you options to edit and work with screensets but also offers a clear indicator from the menu bar of which screenset you're currently working in.

The window combinations can be carefully thought out to maximize both your speed and screen space. For example, most users will keep Screenset 1 as the default Arrange window. However, whatever combination of windows you choose for each Screenset is entirely up to you. One example some users use is to follow the window keyboard commands as a guide; see Window from the main Logic menu. Doing this means that you can get to navigate to the Screenset which focuses on the task you need. For example, Screenset 8 could be set up to be the Environment, or the audio mixer within the Environment window, which we'll cover later in this chapter.

Screensets really come into their own with two monitors. As we've already mentioned, the new Arrange window is a very comprehensive space for achieving most things in Logic, although there will be particular times when you'll want to see additional things, perhaps in larger windows or at the same time as something else. As such, screensets could allow you, if you so desired, to manage the second monitor's content, whilst leaving the Arrange window on the main monitor. Additionally, Screensets can also permit you to choose which editor or list is shown within the Arrange window. For example, one screenset could show the Sample Editor, whilst another the Markers list.

Should you wish to see the mixer and the Arrange window as in the picture opposite, then this is absolutely fine. Although there is another more integrated way of bringing up the mixer using the Environment window, Logic allows for the Environment window to be loaded up as a floating window by

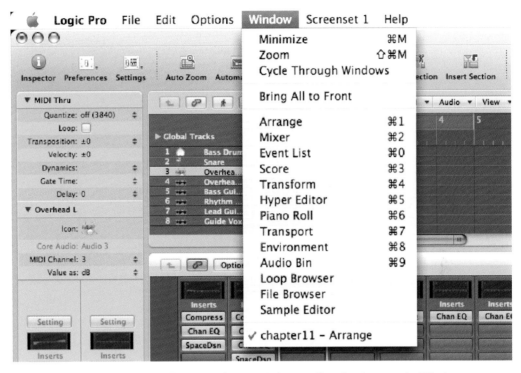

To access all the differing windows Logic has to offer, simply go to the Window menu. This is the menu from which Screensets can be also managed.

Screensets can be really useful when you need to arrange windows to suit your work. Screensets really come into their own when working with two monitors.

pressing Alt as you go to Window > Environment. Before you do this, open a normal Environment window by going to Window > Environment, and change its view to "Mixer" using the arrow pointing down in the top left-hand side of the screen.

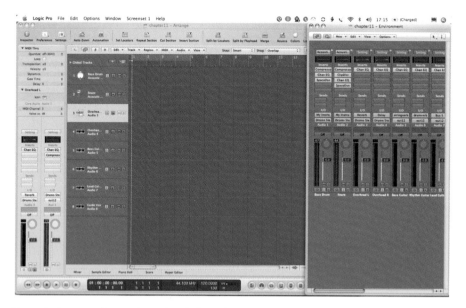

Two windows can be made to sit side by side using Screensets as shown here, but it is also possible to create a floating Environment window which can be linked to your actions on the Arrange area.

With this new floating mixer window, shown below, it is possible to quickly navigate each track on the Arrange window and the mixer will follow as though the link control was on, making it an extremely flexible and powerful way of working efficiently whether on a small, wide, or double-screen setup. To enable this, you need to ensure that the link icon is enabled in the main

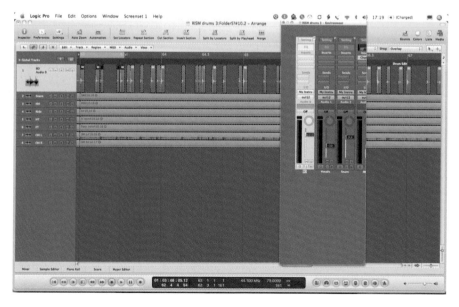

Floating the mixer can be useful to see different elements of the mixer such as the master bus to those elements automatically shown in the Inspector.

Environment window before making it float. This might seem a duplication to the Inspector's expanded feature showing two mixer elements, although you can perhaps be more dependent on the number of monitors you have. The alternative use for this is shown below, where the link has been disabled and we have focused the floating environment window to the master outputs. Now, directly from the Arrange window, we operate the focused track's mixer element in addition to the master bus.

Immediately there are nine screenset levels that can be accessed from the numeric keypad. However, Logic allows you to access and make use of up screensets up to "99." To key these screeensets in you need to press Control for the first digit and type the number of the corresponding screenset in. So for 34, simply press Control+3 and then 4, both on the numeric keypad.

At times you may receive a file you have been asked to mix, which will have different screensets and perhaps different key command settings. It is important that your workflow is not jeopardized and that you can work both efficiently and productively with your preferences. To import project settings from another Logic File, go to File > Project Settings > Import Settings.

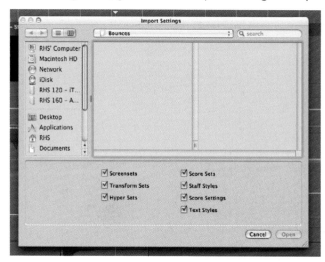

It is possible to import your project working preferences from another Logic File using the Import Settings feature.

Link modes

When you create multi-windowed screensets it is often a good idea to use one of the link modes to connect them all together when editing. As we have mentioned earlier in the book, Link allows you to click on another region and its contents to be displayed in the accompanying editor. This is highly beneficial and ensures that you can work really fast whilst keeping all the windows you need open. If this feature did not exist you'd need to close the window and perhaps double-click on the new region each time. Laborious!

Link has two discrete modes. The first is Same Link Level (pink icon) which means that whatever item you click on, whether that be a region, or the note, the linked windows will follow the same level. Confusing isn't it? Well this can be useful if you have a number of editors open at any one time. A good way to see this working is by opening a separate events list window (Apple+0) and, within the Arrange area, selecting a MIDI region. The events list should show a list of the regions. However, if you press "P" to open up the Piano Roll Editor and then select on the MIDI notes, the Events List will change to the same level or class of data, in this case a MIDI note within all the notes in the region.

Content Link (Yellow Icon) differs in that you always see the content of the region. So in the example above, if we clicked on the region in the Arrange area, the events list represented a list of the regions, but if in Content Link Mode we selected a MIDI note in the Piano Roll then the MIDI note would be shown in the events list. With Content Link we can simply click on the MIDI region in the Arrange area to immediately see its contents in the events list, therefore seeing the content level without having to select it separately in the Piano Roll.

These differences are ever so subtle and frankly it is one of those things that if the windows are not linking as you need them to, then simply Control+click the link icon, change the mode, and see if that works the way you want it to. Nevertheless, the Link mode is a key feature of Logic as it does ensure that screensets can operate fluidly without the need to shut an editor down to then double-click an alternative region to look at its notes!

11.4 Key commands

Keyboard shortcuts are a useful way of improving the workflow with whatever software you are using from a word processor to an internet browser. Logic,

Key commands are an important feature of any piece of software to increase your productivity. Logic is no exception allowing you to carve out a key command for literally every feature.

of course, is no different, and as you'd expect there are so many of them covering a vast number of the features. Throughout this book, we give you the default key commands that relate to what we have covered. Most of these commands can be learnt by noting the writing to the right of the command in one of the menus, but the whole range can be viewed in their own dedicated Key Commands viewer (Logic Pro > Preferences > Key Commands... or Alt+K).

Searching for Key Commands is easy by using the spotlight-like search bar on the top right of the pane. In here simply type in the command you wish to find the key command for and the possible results will appear under one of the main headings in the command list. Under each of these 17 headings, are all the key commands available for use with Logic. Not all of these commands will have a shortcut associated with them as there are simply too many. But as with all things in Logic, they too can be optimized or altered to suit your working needs.

Once you have selected the instruction you wish to create a key command for, select it and choose to either "Learn by Key Label" or "Learn by Key Position" followed by the key combination you wish to use. There are subtle differences between these buttons. Logic allows you to discern between using the numerical keypad and using the numbers on the main part of the keyboard. In this way you could choose to assign two different commands to number 9, one would be on the main keyboard and the other on the numerical keypad. Therefore learning by Key Label means that using the number 1 will call up the same key command whether it is on the numerical keypad or above the qwerty. On the other hand, the Key Position option will allow you to discern between the two differing positions, thus giving you further key command possibilities.

There will perhaps come a time when you will develop your own key commands and see your productivity suit the way in which you work. Your templates will already pre-load your key commands with your work if you saved them, but there may come a time when you have to work on an imported project such as a mix that someone else put together. In this instance it is really important that you save a set of your personalized key commands.

There will be times when you may need to impose your special and personalized changes to another's Logic session and will need to have your key commands ready. To do this we need to export our key commands. To do this simply press Alt+K from anywhere in Logic to bring up the Key Commands dialogue. Within the options menu is a command called Export Key Commands which will allow you to save your key commands. These can then easily be exported into another copy of Logic using the Import Key Commands instruction in the same menu. These key commands can also be saved to the clipboard for use as a reference, or for revision. It may even be worth doing this whilst you're learning to use these commands.

If, however, you have learnt Logic's default Key Command set and import another person's project, you might be at a loss as to why so many of them did not work. In this case, it would be sensible to choose the option of "Initialize All Key Commands," which will reset the commands to the factory standard.

11.5 The Environment

Logic's Environment window has an interesting reputation to the non-Logic using world. It has always held an almost mythical-like status, where many users have not wished to delve for fear it would mess their whole project up. Whilst the Environment is perhaps less used these days now that MIDI is not the main method of music creation within the application, it is still an important tool that allows the user to experiment and perhaps optimize their creative output. The Environment window is thus the route for information into and out of Logic and as such is something that is worth considering especially when working with MIDI. It is here that the virtual representation of your studio can be created, managed and manipulated. In this section we'll give a quick overview of its core functions and its possibilities. Where you take it will be up to your creativity.

Layers and windows

The Environment window itself could be simply looked upon as an area to organize your MIDI inputs and outputs and how they all connect together. However, there are some more angles, or layers, to it which we'll explore a little here such as its Audio Mixer and Global Object layers. In essence the Environment should show all the connections to and from Logic, both physical and internal.

The first layer in the list is the All Objects list which shows you all the connections to Logic currently established. This includes MIDI, Audio and internal connections such as ReWire. The next layer is Global Objects which allows you to specify any objects that are common to each layer. The Click and Ports layer refers to how the click track (metronome) is managed within the MIDI environment, and how the Ports connected to each MIDI input whether that be physical through an interface (port) or the Caps Lock Keyboard. The MIDI Instr., or MIDI Instruments layer is to allow you to specify what each instrument is and to manage its operation from within Logic. The last is the mixer layer which allows you to create a customizable mixer layout to suit your way of working.

New layers can be created for different purposes by selecting "Create Layer" from the menu. It might be prudent to create perhaps an additional layer if you wished to use the Environment to control an external MIDI synthesizer using something we call a mixermap, or a controller. This map could allow a physical display of the controls of that MIDI synth within a new Environment

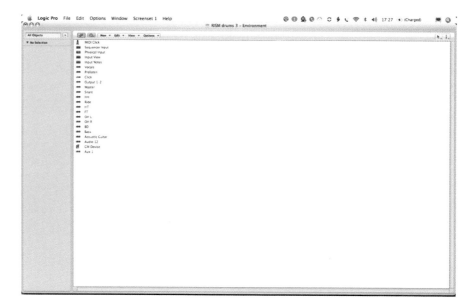

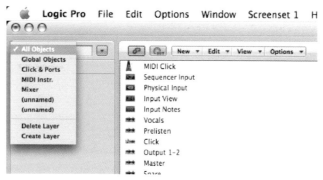

The Environment has a number of different layers which can be selected using the menu at the top left-hand side of the window.

layer. The benefit is here that any movements you make to the synth within Logic can be recalled and automated. What you choose to use additional layers will depend on the way you work. Additional mixer layers might be an idea, one for the channels and another for auxiliary tracks, for example.

An additional dimension to the "layout" of your windows is the opportunity to make the Environment window a Frameless Floating Window by selecting it from the View menu. If you press link on the Environment window before making it frameless, it will follow the regions you select as you move around the project. As the Environment window can be used for so many specialized and personalized facilities, it might be necessary to have it open on top of other windows. In this instance create a frameless window and place it in your screenset next to your Arrange window or however you see fit for your session management.

Objects

Each element within a layer is known as an Object and hence we see the concept of the Environment. The Objects and their connections make up how Logic operates. To see the types of Object available, go to the New menu

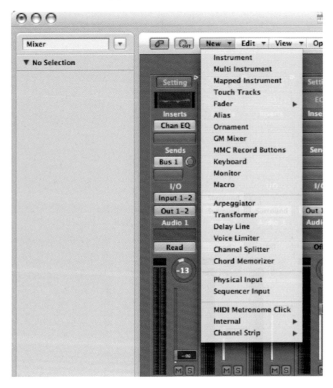

There are a considerable number of Environment Objects available for you to choose to control Logic in a flexible manner.

within the Environment window where you will be greeted by a long list. The list is grouped into four groups. The first is the objects themselves; the second is a collection of processors such as the Arpeggiator or Transformer. The third collection comprises simply two inputs: Physical Input and Sequencer Input. The physical input refers to the MIDI input from an external device, perhaps a master keyboard, whilst the Sequencer input refers to what Logic "gets." In other words, one could place a process such as the Arpeggiator between the Physical Input and the Sequencer Input. The last collection refers to Audio and external functionality such as new Internal connections which include ReWire, and Audio objects such as auxiliary tracks.

Managing objects is simple as they can be moved, copied and edited in a way similar to that of regions on the Arrange area. The Environment window also has an inspector just as the Arrange window and can be toggled in and out of view using the "I" key. Selecting on an object will reveal some editable features within the Inspector.

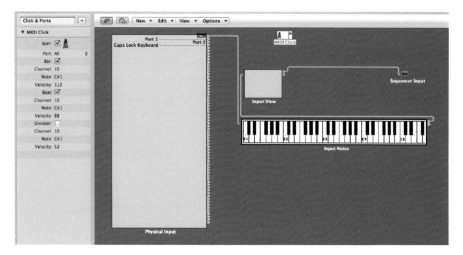

The Inspector is an important aspect of the Environment window and can be toggled in and out of view using the "I" key.

It is quite easy to create some unique layouts such as mixers which would benefit from being presented in a structured and uniform manner. When creating an environment it is easy for the objects to not quite line up. As with many graphics applications, Logic's Environment also includes a snap to grid-type facility called Snap Positions (in the Environment, View > Snap Positions). This will ensure that the objects you place can be easily managed and look smart!

Cables

Objects need to be connected to operate in many instances. Therefore, Logic provides "cables" to connect one object to another and can therefore be extremely flexible.

If you navigate to the Click and Ports layer, you'll notice the Physical Inputs box has a number of white triangular arrows running down the right-hand side. These are connected to some writing that indicates which physical MIDI input they refer, such as Port 1, etc. Alternatively at the top of this box is a SUM of all the physical inputs (which can be really useful if your master keyboard is away from your computer as you could employ the Caps Lock Keyboard or another smaller keyboard) and either signals will be passed into Logic.

It is likely that the SUM output of the Physical Input box has by default been connected to an Input Notes keyboard on screen which allows you to click and preview notes as though you were playing a keyboard. This is usually automatically connected to what is labelled as the Input View, but is actually known as a Monitor Object. This shows you the MIDI information as you play the keyboard, which is usually a Note On message followed by a Note Off for each key pressed. MIDI information then travels through the cable connected to the Sequencer Input which connects to your chosen track on the Arrange window.

As a default, this is great as it allows you to see how the Environment connects and it is at this point that you perhaps might wish to experiment a little. For example, try creating an Arpeggiator and another Monitor Object. The idea here is to place the Arpeggiator after the original Input View Monitor to create a new musical arrangement. This could then be outputted to another Monitor Object to see what changes have been made to your input. With this arrangement of objects, anything played in on any physical inputs will be subjected to an arpeggiator.

The Arpeggiator will run at the tempo of the project, and its response and range can be altered by selecting it in the Environment and visiting the Inspector to the left of the screen.

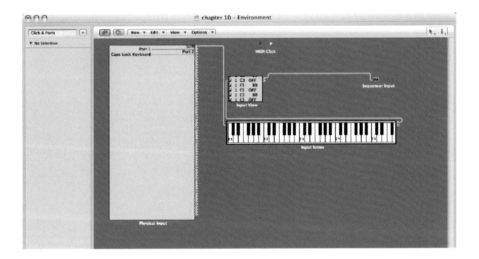

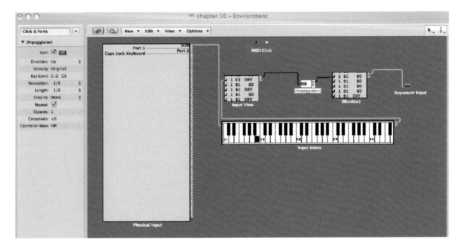

Using the Environment to create unique and interesting creative aspects to your work is easy. In the examples here, we have placed an Arpeggiator across any MIDI input to Logic.

As Click and Ports generally refers to the input side of Logic there is so much experimentation we can do here. There are obviously many more options: we could try working on the output of Logic which can be quickly demonstrated by moving the mixer layer and adding in a new Arpeggiator into the environment. Connect this to a software instrument channel at the top of the fader. This new Arpeggiator object on the mixer will now be available on the Arrange window. Create a new track and Control+click the track header to open up the Reassign Track Object menu. From here select the Mixer sub-menu and you should see the word Arpeggiator written in. This will now allow you to record and play only the source notes which will then be arpeggiated each time the sequence is played.

It is quite possible that the cables on an environment may get so frenetic that you'll not be in a position to see the objects! Within each layer is the option to view or hide the cables thus allowing you to see what is going on. Simply go to the View menu and select cables. This will hide them until you next need them. To make the workflow quicker, you can simply press Control+C to toggle them in and out of view.

Another dimension is the ability to colour your environment cables by clicking on the source, i.e. the device that is outputting to the cable, and pressing Alt+C or View > Colours. Change the colour as necessary to enable quick access to the appropriate element of your environment. Remember these can simply get toggled out of view if it all becomes too much.

Setting up MIDI instruments in the Environment

The MIDI Instruments Layer is an important one as it allows you to specify the MIDI environment and the synths and devices you have attached to your system. For example, if you bring in a new MIDI synth to your setup, you could simply just tell Logic there is a new instrument connected and this will allow you to select it from the track assignments, which is fine, but there is so much more that can be done.

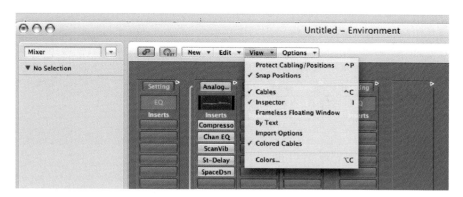

It is easy to view or hide the environment's cables by going to View > Cables.

Creating a new instrument within the MIDI Instruments layer will enable a new instrument to be located within the Reassign Track Object menu meaning

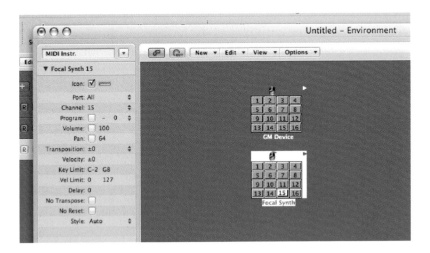

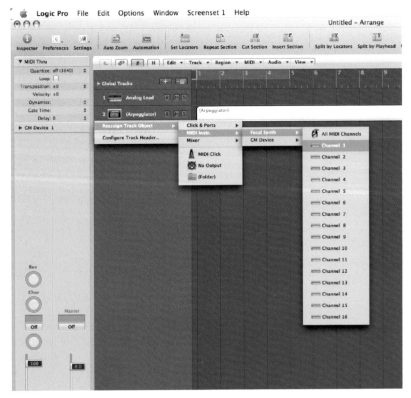

Creating a new Multi-Instrument in the MIDI Instruments layer will allow you to create a new stream of MIDI information intended for this new device.

you can directly select the device and its channel from the Arrange window. To do this we need to create a new Environment Object, and there are two main alternatives for MIDI instruments. The first is to create a new "Instrument" (New > Instrument). This means that this output will go to one MIDI channel from any port. There are some classic examples of MIDI instruments which are not multi-timbral such as the Waldorf Pulse which is a single MIDI channelled monosynth.

For larger multi-timbral synthesizers we'll need to create a "Multi-Instrument." This can allow us to specify quite a lot about the device in question. As you select this, you're confronted by a box with 16 small buttons. These will all be crossed off. This is essentially to save your time as not all multi-timbral synthesizers are able to use all 16 MIDI channels. Therefore you can specify which outputs you wish to use. Another reason for limiting the output of certain channels might be because you may wish to use a channel from the same physical output, if your MIDI outputs are limited, to send to a monosynth such as the pulse on channel 16.

After creating an instrument object, you will need to connect it to an output port. To do this simply select the instrument in question and visit the Inspector to the left of the environment window. Usually directly under the Icon checkbox is a drop down menu which lists all the available physical ports. Simply select the physical MIDI port you have connected the device and you can use the instrument straight away from the Arrange window. This is known in Logic speak as a direct output assignment.

Once you have set the device using the multi-instrument object, you can double-click the object to reveal the Multi-Instrument Window. This immediately shows the instrument names for a General MIDI device that relate to a program change that can be sent by the sequencer. The idea is that you can re-label the instruments within Logic as they relate to their corresponding program change number. Once this has been done, you can choose the instrument name when selecting the Track Assignment in the Arrange window. Whilst this might seem like a long-winded exercise for all your esoteric synths, the operation would only need to be completed once. That level of recall from the Arrange window is an excellent feature and once done can speed up your workflow without leaving your mix position.

The Mapped Instrument Object can be useful when you wish to convert one type of MIDI input data to another. For example, you may have drum kit loaded in an external device which does not follow the General MIDI drum convention. In that instance it would be useful to call up a Mapped Instrument Object which can be double-clicked to open a chart which allows one note to be mapped to another in real time.

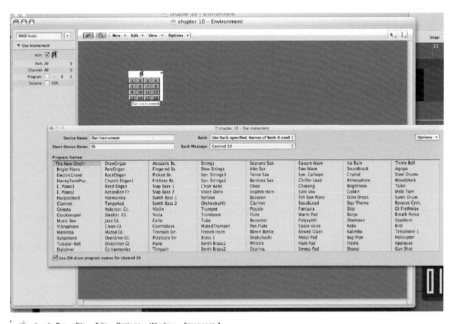

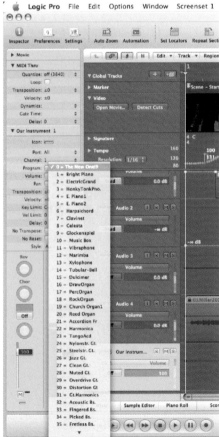

The multi-instrument window allows you to label the instruments or patches to their corresponding program change. This can then be selected from the Program Change menu in the Inspector.

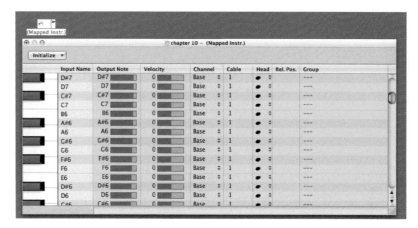

The Mapped Instrument Object converts one set of MIDI notes into another and is primarily useful for mapping drum kit sounds.

Walkthrough

Adding new MIDI devices to the Environment

Step 1:

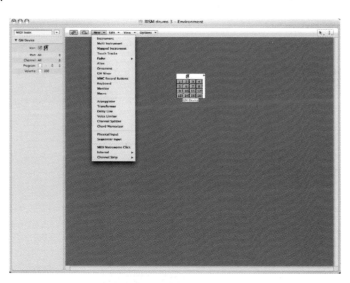

Adding a new MIDI instrument to the Environment will allow you to select it as a distinct item from the Reassign Track Objects menu. First, open up the Environment window, which is Apple+8 or Windows > Environment. With the Environment open, select the layer selection menu to the top left. From this menu select the MIDI Instruments layer abbreviated to "MIDI Instr." This layer should see a General MIDI Device (GM Device) which is sometimes routed to the Quicktime Music Player.

Step 2:

To create a new instrument choose the New menu. There are plenty of options available here, but namely we're concerned with either a new Instrument or Multi-Instrument. An "Instrument" in this case is an instrument which only has one MIDI channel such as synthesiser. The Multi-Instrument alternative is for use with multi-timbral MIDI devices such as a sound module or external sampler. It is important to establish which type your device is before connecting as it will make your selection easier from a track in the Arrange window. When this appears, you will need to route it out of a physical MIDI output. To do this select the instrument object and visit the Inspector on the left to select the port it will output from.

Step 3:

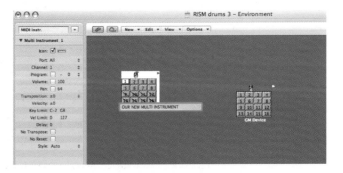

If you have chosen the Multi-Instrument option, a grid with 16 numbers will greet you and they'll all be crossed out. This is so that you can specify how multi-timbral your device is. For example, some synths only can receive 8 channels of MIDI, rather than the full quota of 16. In this instance simply don't deselect the last 8. This will ensure that Logic can only talk to the eight live channels. Whilst you're here, it is really handy if you label the instrument by either selecting the object and clicking on its name in the inspector or selecting the Text Tool from the Click Tools menu. Your device will be labelled and available from the track assignments in the Arrange window.

Live MIDI with the Environment

Earlier we spoke of the Arpeggiator in terms of cabling and the creative effects it can have. The sheer number of objects possible within the Environment window allows for us to design some interesting and important facilities for working with MIDI. Whether that be an Arpeggiator or Multi-Instrument, the two together can make for some interesting outputs. This can of course be extended further if the fancy takes you. In the couple of examples below we look at some aspects which can make Logic a useful tool when working live.

There are some interesting elements to the software which might be worth mentioning here. First is the Chord Memorizer Object whose job in life is to do a more clever version of what many lower end keyboards are renowned for: one note chords. The concept is that you can detail what the input note will be and what chord results from it. To create this simply visit the View menu and click on a Chord Memorizer Object. Next double-click the object itself to reveal two keyboards which allow you to set which chord plays depending on which key would be pressed.

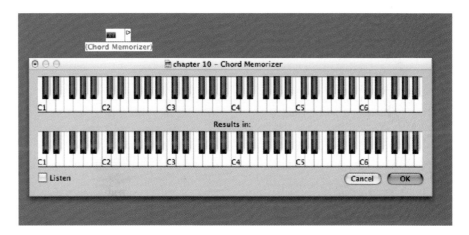

Chords can be memorized to be played using one key using the Chord Memorizer.

The Touch Tracks Object offers the Logic user a range of triggers to be enabled to play certain regions when pressed, like one of the functions of Ableton Live. This is a process which is limited to MIDI regions and Folder Track Regions. Select a Touch Tracks Object from the New menu and double-click it as it comes into view. Another table appears to allow you to assign a region or folder region to certain keys. This can be incredibly flexible when working Live.

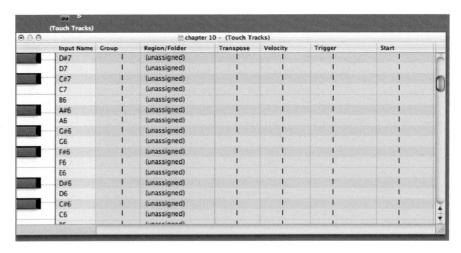

Touch Tracks enables Logic users to work with a range of MIDI regions mapped to various keys. Great for a live performance!

External control with the Environment

One of the most flexible aspects of Logic is its ability to control external equipment which responds to MIDI. Whether to an external analogue or digital mixer that responds to MIDI messages or a synthesizer, Logic can control whatever functionality it allows. This can be extremely flexible to allow you to further control your equipment from the mix position. Any moves you make can be saved with the project as the MIDI can be saved as automation.

To enable this, simply create a new layer which can be renamed as the device you're planning on automating. Within this layer, create a series of objects which reflect the device and the controls you want to manipulate. Each object will have a different function to manage in the destination device. To alter this, we need to select each object and alter the MIDI message type it sends out. In the example on p. 354 we wish this fader to manage a volume; hence, this should be control change 7 on MIDI channel 1. This would need to be repeated for each channel and message for the mixer.

There will, however, be instances where the MIDI data required will need to be somewhat more complex such as System Exclusive (Sys EX) which will be required for the Mixer. For these message types it will be trickier to enable as this takes more knowledge of MIDI than the remit of this book, but is of course possible. If your external synthesizer follows General MIDI, then simply call up Logic's own GM Mixer by creating a new layer and then visiting New > GM Mixer. This might be a good starting point for the creation of any new mixer maps.

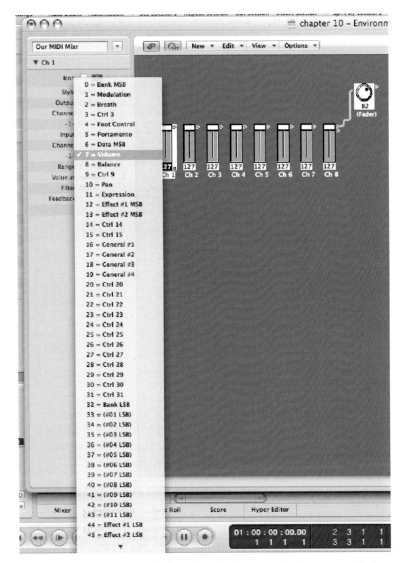

To change the assignment of the fader, simply go to the Inspector and select the type of message and the corresponding number. In this case we should send out a volume message on MIDI channel 1.

When creating mixermap environments to control externally connected MIDI instruments, it is worth noting that these are best designed on the output side of the Environment. By 'output side' we mean by cabling up the faders directly before introducing a physical output using an Instrument Object as can be seen in the diagram below.

There is a bit of a strange logic to this, but doing it this way you immediately take control the device from the mixermap by connecting the fader objects through to the MIDI instrument. Logic immediately records any actions you

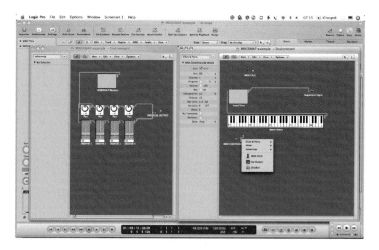

When making Mixermaps, it is advisable to cable your fader objects
directly to a physical MIDI output using an instrument object.

make on the environment page should you wish to without the need to cable
these object to another sequencer input.

Your MIDI device, which might be a mixer as in the environment example
above, may also be able to transmit MIDI commands as you alter its physical controls. If so, these can be in turn fed-back into Logic and be reflected
on the Environment's mixermap providing it is cabled correctly. This loop is
necessary to ensure that both the physical and virtual controls are representative of each other. Fader Objects on the Environment will only be controlled,
though, if the physical device is routed back to the mixermap.

To do all this, simply create a new layer and name it 'Mixermap' or something similar. Create an instrument (New > Instrument) which will act as your
'Physical Output' directly to the MIDI controlled device, in this example a
mixer. It is worth ensuring that the icon checkbox is unchecked for this object
as this does not prevent the icon from showing in the environment (we've
changed it to a MIDI Plug here), but it will no longer show up in the arrange
area when selecting outputs.

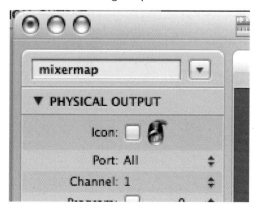

The Icon checkbox does not prevent
the icon beside it being shown in
the Environment itself, but refers
to whether the selected instrument
object appears in the outputs on the
arrange area.

Next select the faders and other object types that best describe how you wish to control the external device. Select each new object and specify its parameters, such as control change number 7 for volume and control change 10 for pan, if it were a General MIDI device. It is worth noting at this point that many MIDI devices use different message types than you'd expect so it's best to visit the corresponding manufacturer's manual. Alternatively, put Logic into record and move the appropriate fader on the MIDI device. Once recorded, simply visit the events list (Window > Events List or Apple + 0) to ascertain what control change message it transmits. Later we'll show another easier way to do this in the Environment window.

In order for this to operate effectively, you need to cable up the faders and other objects to the newly created instrument we've called 'Physical Output'. To the right hand side of each object are some triangles that represent cable 'sockets'. Simply click here and drag the mouse away to show some virtual cables which can be dropped onto other objects to make a connection as in the example above.

These cables then need to connect to the newly created 'Physical Output' to transmit the changes you make on screen to the device. This should work providing the MIDI interface and MIDI device are operating properly.

However we need to ensure that any movement on the physical controls on the MIDI device are reflected on your Environment Mixermap. To do this we need to make a connection between the physical input to the Environment and the Mixermap layer. Firstly it is necessary to add a new monitor object to the newly created mixer. This Monitor object will act as both a way of visualizing the MIDI information as we would in the Events List and as a method of connecting layers together. Connect the Monitor's output to the first mixer object.

We then need to create a new instrument object on the Clicks & Ports environment page if you have not done so already for this MIDI device. In the example above we've named it "MIDI controlled Mixer". If this were a synthesizer or sound module, then this will appear in the Arrange area track lists as usual.

In order to get the Mixermap Environment layer to respond we need to cable this instrument object on the Clicks and Ports layer to the Mixermap itself. Simply select a cable output from the Instrument Object on the Clicks & Ports layer and with both layers open simply drag and drop the cables between layers to the Monitor object. An easier way would be to Alt+Click the output of the Instrument Object to which reveals a menu similar to that found on the track list on the arrange area. This second method will only work, providing you have enabled the 'Icon' check box which relates to the Mixermap's Monitor object.

Providing all the objects are cabled correctly, your newly created Mixermap should offer real flexibility to control devices directly from your Mix position. The open-ended nature of the Environment allows for many more possibilities to control devices within your studio.

11.6 Input/Output (I/O labels)

When working within a studio environment which brings together a fairly big analogue setup using many processors and perhaps a mixing console, there may be too many connections for you to remember which input relates to your Fairchild unit. Like with so many things within Logic, this can of course be labelled up on some physical inputs. To do this, select Options > Audio > I/O Labels. A new window opens showing all the physical inputs and auxiliaries within the system.

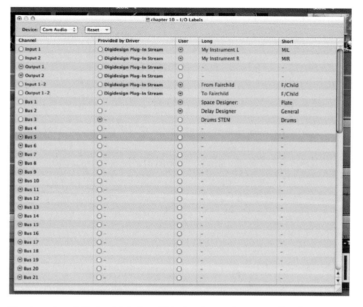

The Input and Output labels within Logic can be named to improve the workflow of the process.

The changes made in the I/O Label dialogue box makes all the difference with respect to quickly recognizing busses in the mixer.

To change the names of connections click either in the Long Name or in the Short name boxes. The Long Name is your description, and should be whittled down to a smaller amount of characters for the short name which will be used more frequently when accessing drop down menus in the mixer.

11.7 Nodes and distributed audio processing

In Chapter 3 we briefly referred to distributed audio processing and the notion of Nodes. Here we discuss how to set them up, get the best out of them and what to do when you want to take the project away without the additional power of the node.

Connecting Nodes to your Logic setup is very straightforward. This is achieved by visiting the Audio Preferences pane and selecting the Nodes tab. There is a small dialogue box which allows you to "Enable Logic Nodes." With this

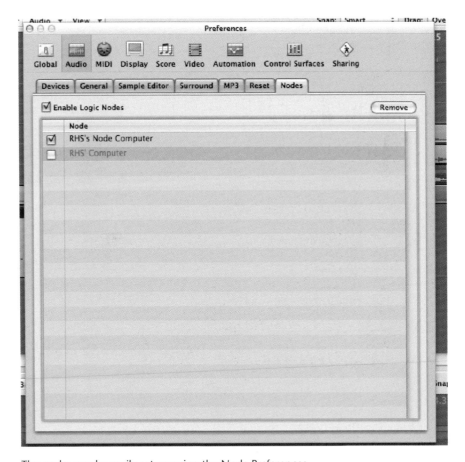

The nodes can be easily set up using the Node Preferences.

ticked, Logic scans for computers connected to the host which have the Logic Node application running. It is not necessary to have Logic Pro installed, only the Logic Node program loaded on the slaved machine. Simply choose the computer you wish to act as the node and then hey presto, this should allow for communication between Macs.

To make use of the node facility, you need to select which aspects of your production are handled externally by the node, or internally by the host computer. This selection is made possible through Track Node Checkboxes which can be revealed by visiting View > Configure Track Header.

The dialogue box (below) appears offering you to customize the way in which you see the track's header, which is the area to the left of the arrangement and

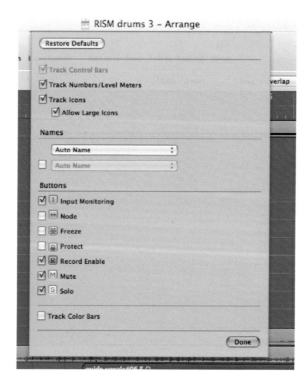

The Track Header Preferences allow for many changes including the Node button asking whether or not a particular channel is to be managed using a Node.

immediately to the right of the Inspector. Within the Configure Track Header dialogue box, the options to change the buttons on offer are selectable. Some useful buttons lurk here such as the Freeze Track icon and in this instance the Node button.

A good way of keeping an eye on how your processing is distributed is to call up the CPU monitor by pressing Alt+X. A small floating window will appear with some lanes. The audio bar has perhaps more than one lane in it which represents the number of processors within your Mac, whilst the Disk I/O shows the data through-put to your hard drive. As a node is added it expands to show the node's CPU load, also allowing you to make a definite assessment of the project's DSP distribution.

Index